Understanding Communications Systems

Written By: Don L. Cannon, Ph.D
Assoc. Prof. of Electrical Engineering
University of Texas at Arlington
Staff Consultant, Texas Instruments Learning Center

Gerald Luecke, MSEE
Mgr. Technical Products Development
Texas Instruments Learning Center

A DIVISION OF TANDY CORPORATION
FT. WORTH, TEXAS 76102

This book was developed by:

The Staff of the Texas Instruments Learning Center
P.O. Box 225012 MS-54
Dallas, Texas 75265

With Contributions By:

Tim Shirey

Appreciation is expressed to Ross Johnson and Dave Woody for their valuable comments.

Artwork and Layout By:

Plunk & Associates

ISBN 0-89512-035-6
Library of Congress Catalog Number: 79-92683

IMPORTANT:
Texas Instruments makes no warranty, either express or implied, including but not limited to any implied warranties of merchantability and fitness for a particular purpose, regarding these materials and makes such materials available solely on an "as-is" basis.

In no event shall Texas Instruments be liable to anyone for special, collateral, incidental, or consequential damages in connection with or arising out of the purchase or use of these materials and the sole and exclusive liability to Texas Instruments, regardless of the form of action, shall not exceed the purchase price of this book.

Table of Contents

Preface

We all are involved in communicating. It may take many forms. A picture, the spoken word, a glance, a particular body movement, a hand-sign, a printed symbol – these are some of the forms. Electronic techniques are contributing to systems that make it easier to communicate information from one place to another. If you are at all interested in how this is being done this book is for you.

A knowledge of electricity and how electronic circuits work would be beneficial – but not necessary. Basic concepts, basic fundamentals are developed from the beginning to help the reader understand how information can be handled as electrical signals, and how these signals are converted from a form used by humans to communicate (sight, sound or touch) into electrical signals that are used to transport the information to remote locations throughout the world.

Whether a person wants to have an overview knowledge of telegraph, telephone, radio, television, facsimile, computer or satellite communications systems or wants a good bit more detail, this book should provide that understanding. An understanding that can serve as a stepping stone to more study on the details of anyone of the subjects.

Many advances have been made in the type and kind of information that is available to all of us. Electronic communication has impacted this significantly. Business management, transactions, results all depend on a wide variety of these systems. More and more individuals will be affected by the advances that are here and will be coming in the future. Whether it be news, personal business, social functions, entertainment, or sports, communications and communications systems will be involved.

This book, like the others in the *Understanding Series*™, is designed to build understanding step by step. For this reason don't skip around in the book. Read each chapter one at a time, especially the first five chapters. Try to fully understand the contents of a chapter before going on to the next one. Each chapter moves quickly to a higher level of understanding.

Quizzes are provided at the end of each chapter to review the basic concepts presented in the chapter.

A glossary and index are provided to aid in using and understanding the material and finding the subjects of interest.

The advent of the integrated circuit has provided the capability to put into a very small space highly reliable, low-power, high performance, amazingly functional electronic circuits. The resultant advances of electronics have improved and will improve the quality, safety and variety of life for individuals and for society. We hope this book helps you to understand why and to appreciate the continuing progress.

D. Cannon

G. Luecke

The World of Communications

A SCENARIO

The year is 1840. Your mind projects you back to the open prairies of the West and the Southwest. You're on horseback and you are traveling through this wide plain forming a valley between high hills on each side. To some they may be mountains, but you have seen ones much higher. Scanning the scene you notice large puffs of smoke rising from one of the highest points on one of the hills to the right of the plain. The puffs continue rising until there are several in a row seeming to form a code. Then they stop. You wait, wondering what happens next. As if on key, gradually rising from a high point on the hills to the left of the plain, similar puffs of smoke start into the air. They continue until a string can be seen rising together above the hill. Then they stop. Is that all? Is that the end? No, now the puffs begin on the right again. Another string is seen rising in the air. The events continue in the same pattern – first a string of puffs on the right, then a string of puffs on the left. Straining your eyes and shading them from the sun, the outline of several people can be seen below each string of smoke puffs. Suddenly you realize that these people are sending messages. The string of puffs of smoke are a code and the code is understood by the people on each hill.

As you continue to ride, you arrive at the base of the hill to the right. Looking up you see that Indians have built a large fire, and are throwing tree branches heavy with green leaves and undergrowth onto the fire to produce the smoke. Four of them hold what looks like a blanket over the fire and then pull it away to make the puffs. You come to realize that you are witnessing an American Indian long distance communications system. It has all the integral parts of the modern day communications system. The transmission is through the air, in the space between the hills and in all directions around them as far as the eye can see. It is line of sight transmission; it has a carrier of the information; it has a code; it has a transmitter and it has a receiver.

The Indians, the fire, the blanket, and the smoke are the transmitter. The eyes of the people are the receiver. The light waves are the carriers of the information; the code is the information. Light waves transmit the information on the carrier from the transmitter to the receiver. Considering all the parts together, they form a fairly sophisticated communications system. Information is being communicated from one location to another remote location a long distance away – through this system the Indians are communicating with each other.

ABOUT THIS BOOK AND THIS CHAPTER

Today, communication is an integral part of everyone's life. Much of the time spent every waking day is spent communicating. Family relations, education, government, business and other organizational activity is totally dependent on communications. Communications is such a commonplace activity that most people take it for granted. Yet without communications most modern human activity would come to a stop and cease to exist. To a very great extent the success of almost every human activity and organization is highly dependent on how available communications methods and techniques are effectively utilized. It is for this reason that it is so important for everyone (except maybe people that want to be or need to be completely isolated) to thoroughly understand what the various methods of communication available to them can and cannot do. It is the purpose of this book to provide a firm foundation of the concepts involved in modern communications systems so that this understanding will be possible. It is the purpose of this chapter to start the understanding by making clear what the words "communications" and "communications systems" mean.

WHAT IS COMMUNICATION?

Since communication is such a commonplace activity, it would seem that the answer to this question would be obvious and quite simple. Most readers should have a good idea of what communication is through their experience with it in their everyday lives. However, if one tries to define communication at very basic physical or psychological levels, the answer to this question can become very involved indeed. In fact, entire volumes have been written on the physical, theoretical, and psychological aspects of communication. It is possible to get an advanced college degree in any one of these areas of communications. Fortunately, it is not necessary or even desirable for everyone to get that involved in communication theory. It is quite sufficient to understand the fundamentals of the subject, and this is the purpose of the next few sections of this book, beginning with some basic definitions.

DEFINITION OF COMMUNICATION

Communication is the transfer of meaningful information from one location (the sender, source, or originator) to a second location (the destination or receiver). This definition is a basic one that only requires that the term information be defined to complete its meaning, Another name for the sender is the transmitter. In this book and in much of electronic communications systems terminology, the names transmitter and receiver are used almost exclusively. The physical path over which the information flows from transmitter to receiver is called the transmission link or the channel. These terms are diagrammed in *Figure 1-1.* This type of drawing is called a block diagram and is used to illustrate how the various parts work together in electronic systems. In the case of communications systems, the block diagram summarizes the flow of information in the system.

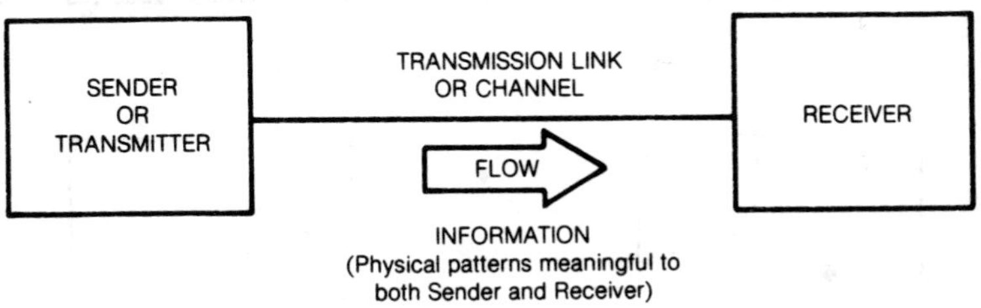

Figure 1-1. *Communications System Diagram.*

DEFINITION OF INFORMATION

Information is a physical pattern that has been assigned a commonly understood meaning. The pattern must be unique (separate and distinct), capable of being sent by the transmitter, and capable of being detected and understood by the receiver. If the information is being exchanged between human communicators, it generally is transmitted in the form of sound, light, or texture patterns in a manner that can be detected by the primary human senses of hearing, sight, and feel. The receiver assumes that no information is being communicated if no recognizable patterns are being received. For example, in *Figure 1-2,* if a person looks at a blank piece of paper or a sheet of paper with just a slash through it *(Figure 1-2a),* that person would most likely assume that no communication was intended. There is no pattern on the paper that is really meaningful to the person. On the other hand, a drawing of a horse, however crudely done, or the name horse spelled out in a language understood by the person *(Figure 1-2b)* would communicate the concept of horse to the person.

The communication might still not be complete or meaningful unless the transmitter or writer and the receiver or reader both agree that seeing the symbol or the name horse meant some complete idea. Other symbols might mean a command, or a warning or a reminder. Whether complete or not, information, or at least part of a communication, has been sent or transferred or transmitted from one person to another.

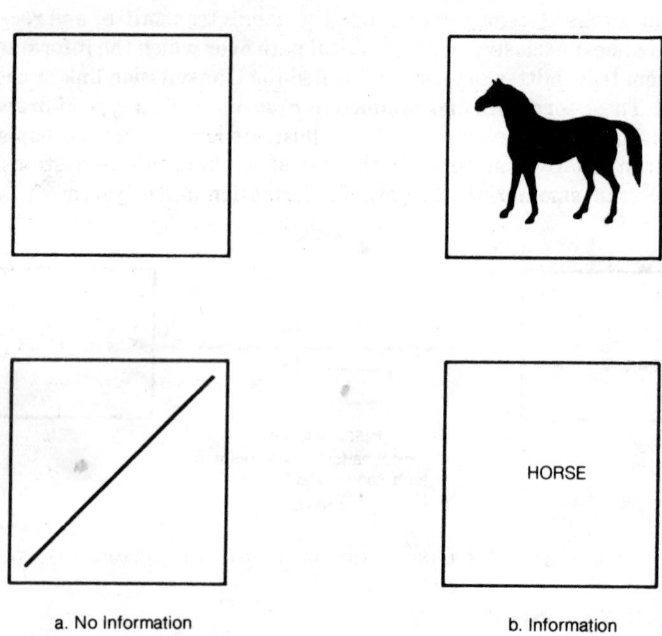

Figure 1-2. *Patterns of Information.*

INFORMATION CODES

If a complete command or warning or other concept has been assigned to a symbol, the information has been encoded. That is, the transmitter has used the code symbol (in the above example, the horse) to transmit an entire message or command. The receiver must interpret or decode this message to understand its meaning.

Generally, all communications systems, particularly electronic communication, use some form of coding and decoding in sending the information from the source to the destination; therefore, a typical block diagram of such a system containing encoders and decoders is shown in *Figure 1-3.* There are two different ways in which information is encoded in communications systems.

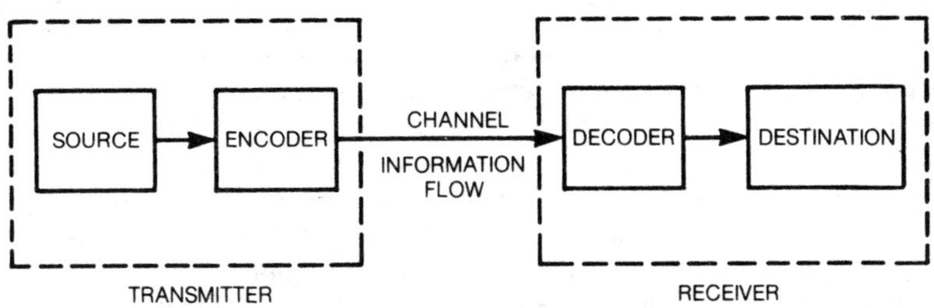

Figure 1-3. *Communication System Block Diagram.*

In almost all cases the information must be placed in a physical form that can be transmitted and received. The encoder performs the conversion so the information can be transmitted. The decoder performs the opposite conversion. It takes the received information in the physical form or pattern as transmitted and converts it into a form suitable for use by the person or machine receiving the information.

Using a symbol or picture, such as the picture of the horse, is a different case. Here the sender and receiver have agreed that the specific physical pattern (the picture of the horse) has a complex meaning (that of a horse). So they don't really need encoding and decoding. The purpose of the encoding/decoding conversions in the system is either for efficiency (a long message can be sent with a single symbol) or for secrecy (a symbol is sent that has meaning only to the sender and receiver.).

Even simple and direct communcations may involve either or both of these types of coding of information. When one person talks to another, the vocal chords of the speaker encode the information in the brain of the speaker into sound waves that can travel over the distance between the two. Similarly, the ear of the listener converts the patterns in the sound waves to electrical patterns meaningful to the brain of the listener. This would be the encoding and decoding of the information into a physical form for transmission over the transmission link or communications channel – in this case, the air between the two people. If the two people were to speak in code words that have meaning only to them they are sending and receiving information and keeping it secret at the same time. Both types of coding are used in electronic communications and will be covered throughout this book.

BASIC COMMUNICATIONS SYSTEM PARAMETERS

Rate of Information Transfer

Before leaving the general terms and definitions of communications systems, other general system requirements of specifications will be examined. *Figure 1-4* lists several of these. All of these requirements are interrelated. They depend on each other, and for a given system the order of importance changes. In most cases, the transfer rate of information is the most important. The amount of information that must be communicated from one person to another in a certain amount of time determines the rate of information transfer. The rate of information transfer will determine the physical form and techniques used to transmit and receive the information and, therefore, determine the way the system is designed and constructed. So frequently a prime system parameter sets the system, and the other requirements fall into place behind it. When information transfer rates are so low that any type system will meet the requirements, then this requirement moves to the bottom of the list and one of the others takes highest priority.

Rate of Information Transfer
Reliability of Information Transfer
Convenience
Cost of transferring information — money and energy

Figure 1-4. *Important Communications Systems Parameters.*

Reliability

Users must be able to depend on a communications system. It must work when needed and transmit and receive information without errors.

The reliability of a communications system can be affected by several factors. First, the system itself should almost never break down or fail to operate. This can be accomplished through the use of modern electronic components in building the system, especially integrated circuits. Secondly, the encoding or decoding conversions, if not designed properly, can cause errors in the information communicated. A careful design of the encoding and decoding devices can avoid this source of error or at least minimize it. Finally, the presence of noise in the communications channel can mask the true signal at the receiver and prevent the receiver from providing an accurate replica of the information transmitted.

Examples of the system noise are as varied as there are communications systems. On a windy day the Indian smoke signals are swept away quickly preventing accurate transfer of information. Trying to communicate information by talking in a rock crushing plant is just about hopeless. Writing messages in poor handwriting or on dirty or smudged paper will mask the message from the reader. Electrical storms, rain, hail, and sleet can play havoc with accurate electronic communication. There is always some noise in the information transfer process. The systems designer and user must make sure that the encoding and decoding process and the selection of the communications channel are such that the effects of this noise are minimized.

Convenience

Many communications systems not only must be reliable, but they must be convenient in order to be effective and efficient. Contrast the telephone and television of today with the smoke signals of the Indians. The television and telephone bring into the living room worldwide news, educational materials and entertainment. Within minutes personal conversations can be held with another person halfway around the world, in real time, and in voices and sounds that are familiar. At best the smoke signals carry their message for 10-20 miles, requiring just the right time, place and conditions before a message can be sent.

Cost

The cost of a system interacts with and relates to each of the other requirements. Obviously, the user always wants the most performance at the least cost, with good reliability and convenience. In many cases, cost is the most important factor, while in other cases it is only one of the deciding factors.

The question of cost must be posed not only in dollars, but in terms of energy and raw materials as well. While a sound warning signal could be generated by exploding an atomic bomb, the cost in terms of dollars, the amount of energy used, and the loss of property would be astronomical. A siren is much more cost effective. The telephone is even more cost effective considering what it can do. It can sound warnings, carry business and personal conversations, couple together computers and a lot more. Its system is much more expensive than the siren, but its overall cost in terms of its overall capability to each user is much more cost effective. In this case, the key to determining the effectiveness of the cost of a system is in terms of its cost per unit of information transferred. Effective comparisions of the cost of alternative system approaches can be made in this fashion.

Electronic communications systems must meet the same requirements. This book will cover the special features that make them so attractive in meeting these requirements. But before getting into some of the basic features of electronic communications, a brief history and discussion of the importance of communication to our human evolution will be useful to place the subject in better perspective.

WHY IS COMMUNICATION IMPORTANT?

Let's examine *Figure 1-5a*. Much of the human race's progress in all areas has been influenced relatively directly by the ability to communicate. During primitive ages, man was limited to communications through gestures, facial and body expressions, and simple verbal utterings. His range of communication was limited to his immediate vicinity and his rate of information transfer was very low. As a result, the early social groups were autonomous, with small groups of people banding together in packs, with little interaction between such groups, except for fighting over territory and food supplies. As man's ability to communicate improved, so did his social and economic situation. By the early civilizations, man's communicative ability increased to the use of spoken language, written language, and the visual and musical arts. These early civilizations had relatively complex social and political structures as well as somewhat extensive trade and commerce activities. Man had advanced through the family and tribal units to entire regional empires. With couriers to deliver messages and with written and spoken languages available, the range of communication extended over hundreds of miles, though the time to deliver these messages might extend into weeks. As a result, the rate of information transfer was still low, though many orders of magnitude higher than that of early civilization.

This situation remained relatively constant until the development of the printing press (1440-1460) which made large amounts of information available over a wide range of territory. Delivery of the information was still slow, depending on the relatively slow transportation systems available. However, because of much more current commercial, governmental and scientific knowledge there was a noticeable impact on the economic and social evolution. The limiting factor on the effectiveness of communication still was the rate and range of transfer of information.

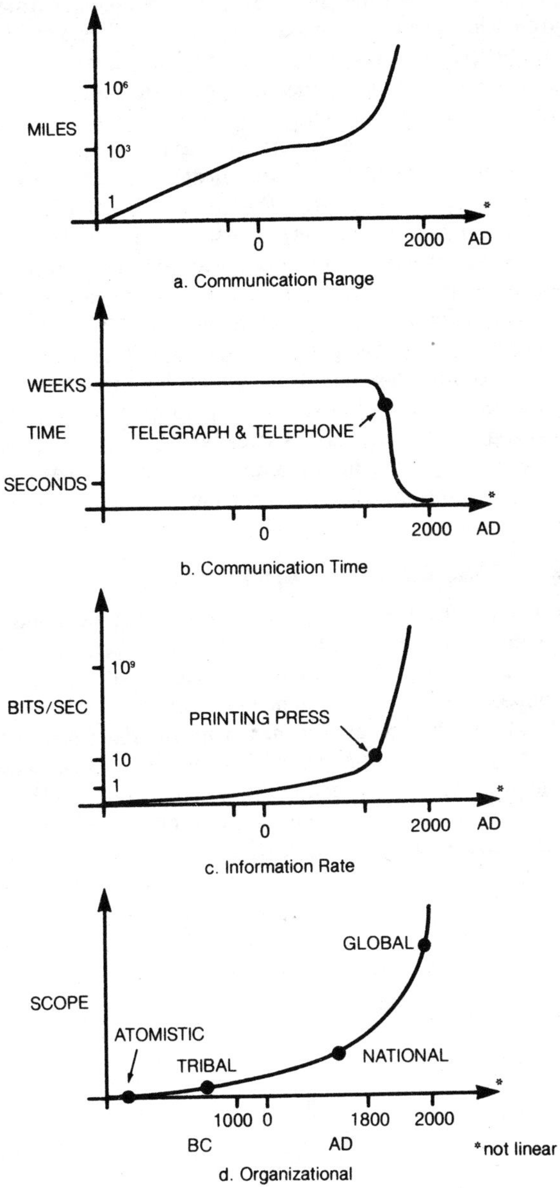

a. Communication Range

b. Communication Time

c. Information Rate

d. Organizational

*not linear

Figure 1-5. *Man's Communications Evolution.*

This limitation was overcome dramatically with the development of the telegraph and telephone (1844-1876). The range over which man could communicate within seconds reached thousands of miles, even linking continents together. From this point, international commerce, communications, and cooperation began to expand rapidly. The subsequent development of radio and then television further accelerated the transition to global interraction of people. Government and business took the lead. With the development of satellite communications systems and computer control of communications networks, the rate and range of communication today allows the global interconnection of telephone, television, and computer systems to provide fast communications to any location on earth and even beyond. It is now within a person's communicative ability to meet almost any information transfer requirement.

Society's progress has gone hand-in-hand with the ability to communicate. As communications systems develop further, it will be of utmost importance that each individual keep up with the capabilities and make the best use of communications alternatives. Future success may depend on how these systems are understood and used. To help understand modern communications methods, let's review historically how man has communicated.

HOW HAVE HUMANS COMMUNICATED?

Most communication in the past has been between one human and another. As a result, information has been sent in a form that is compatible with a person's physical ability to generate patterns and within the capabilities of a person's senses. The primary senses used have been hearing and sight, with some approaches using the sense of touch. The transmitter (the person sending the information) had to generate patterns of light, sound or texture. Thus, communications systems in the past used two broad techniques – visual and aural. The evolution of people's ability to use these techniques is summarized in *Figure 1-6*.

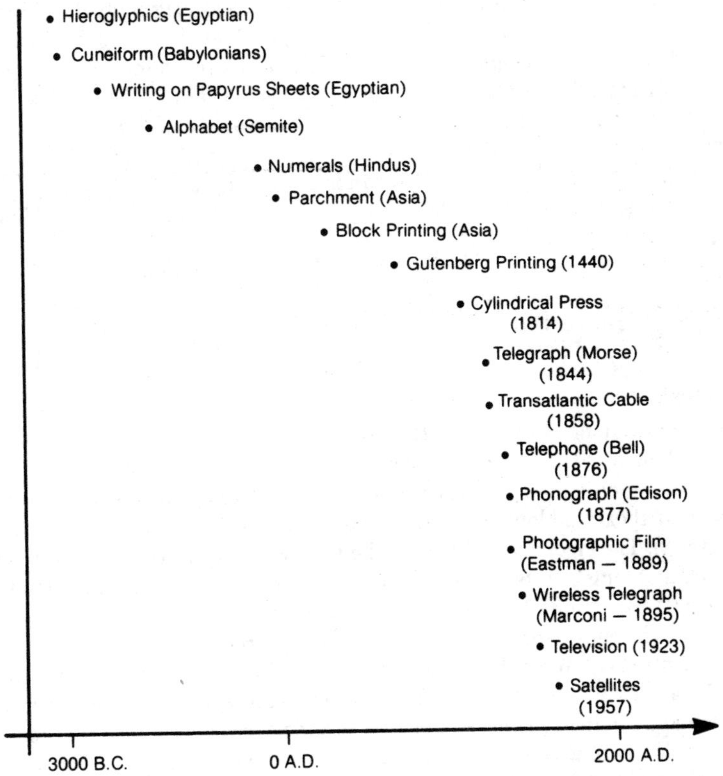

- Hieroglyphics (Egyptian)
- Cuneiform (Babylonians)
- Writing on Papyrus Sheets (Egyptian)
- Alphabet (Semite)
- Numerals (Hindus)
- Parchment (Asia)
- Block Printing (Asia)
- Gutenberg Printing (1440)
- Cylindrical Press (1814)
- Telegraph (Morse) (1844)
- Transatlantic Cable (1858)
- Telephone (Bell) (1876)
- Phonograph (Edison) (1877)
- Photographic Film (Eastman — 1889)
- Wireless Telegraph (Marconi — 1895)
- Television (1923)
- Satellites (1957)

3000 B.C. 0 A.D. 2000 A.D.

Figure 1-6. *Evolution of the Human Race's Communication Techniques.*

COMMUNICATING WITH SOUND PATTERNS

Speech

The most common and convenient method of communicating with sound is human speech. Some say this first started with basic utterings such as growling to express anger or a threat, and developed into sounds to represent specific things or needs. Such sounds were the first elements of a language in the form of simple words. Eventually the vocabulary grew to entire languages. In all of these cases the transmission of the information is in the form of sound waves generated by the speaker's vocal chords. The reception of the information is by the listener's ears or hearing.

Drums

The problem with spoken language as it was originally developed is that it is relatively slow and limited in range. Even with loud yells the range of this form of communication is limited to a few hundred feet, and the rate of transfer to a hundred or more words per minute. The range limitation of the sound communication was overcome early on with the use of drums or other percussion effects to generate impulses of sound energy that could carry messages over thousands of feet. By providing a network of drummers, messages could be relayed over long distances. The problem with this method was that the rate of information transfer remained severely limited. Generally, only simple warning or announcement messages could be sent. Further, a code system was necessary. Some examples might be: three drum beats means danger, or a certain pattern of drum rhythm and tones may mean that friendly visitors are approaching.

Telegraph

The modern version of the drum method is the telegraph developed in 1844. This was the earliest form of electrical communication. At the transmitting end a telegrapher closed a switch or telegraph key in a certain pattern of short and long closures to represent a letter of the alphabet (see Chapter 2). The electrical energy on the wire was sent in the same pattern of short and long bursts. At the receiving end this energy was converted into a pattern of sound clicks that was decoded by a telegrapher at that end. The code used by both transmitter and receiver is the Morse code *(Figure 1-7)*. With this system and the laying of the transoceanic cable in 1858, a person's range of communication expanded to thousands of miles. The message delivery time dropped to seconds over this range, and the information rate was maintained in the 5 to 100 word per minute range. The only problem with the approach was that only skilled telegraphers could actually use the system. Everyone that wanted to send a message had to go through the telegrapher.

Telephone

The development of the telephone in 1876 *(Figure 1-8)* was again an electrical communications system. In this system, the speaker's voice was converted into electrical energy patterns which could be sent over reasonably long distances over wires to a receiver which would convert these energy patterns back into the original sound waves for the listener. This system provided many of the long range communications capabilities of the telegraph, but also had the convenience of using speaking and hearing directly so that everyone could use the system. Its rate of information transfer was limited by the rate of human speech.

TELEGRAPH CHARACTERS

	Morse		Morse
A	• —	T	—
B	— • • •	U	• • —
C	— • — •	V	• • • —
D	— • •	W	• — —
E	•	X	— • • —
F	• • — •	Y	— • — —
G	— — •	Z	— — • •
H	• • • •		
I	• •		
J	• — — —	1	• — — — —
K	— • —	2	• • — — —
L	• — • •	3	• • • — —
M	— —	4	• • • • —
N	— •	5	• • • • •
O	— — —	6	— • • • •
P	• — — •	7	— — • • •
Q	— — • —	8	— — — • •
R	• — •	9	— — — — •
S	• • •	0	— — — — —

Figure 1-7. *International Morse Code.*

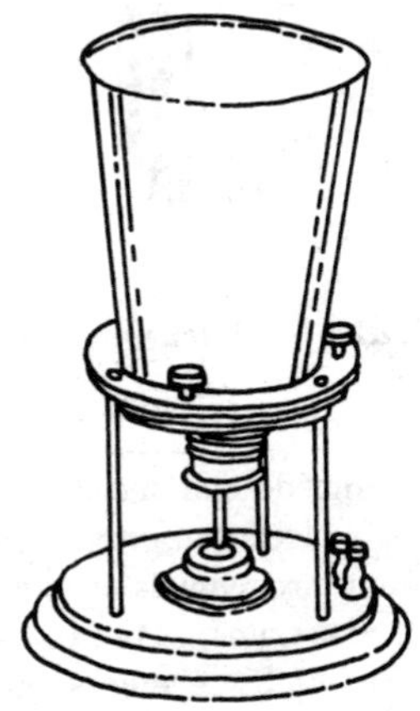

Figure 1-8. *Early Telephone.*

Wireless Telegraph

All that really changed from this point in the area of sound communications was the transmission approach. The development of wireless telegraphy in 1895, the forerunner of modern radio, allowed telegraph communications to occur over long distances without the need for running wire from the transmitter to the receiver. Now messages could be sent to mobile receivers on ships, military units, and remote expeditions. The communications transmission link was electromagnetic waves that radiated from the transmitter location to the receiver location.

Wireless transmission was not practical without the perfection of the vacuum tube shown in *Figure 1-9* which provided the carrier frequencies at the power required in order to convert information to electromagnetic radiation.

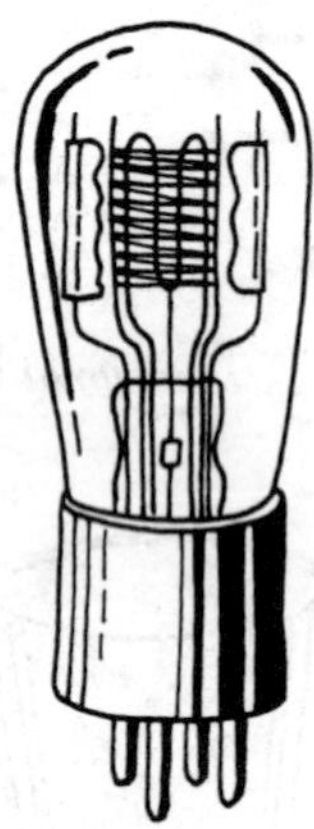

Figure 1-9. *Vacuum Tube.*

Radio

Once man learned to encode and decode the human voice in a form that could be superimposed into electromagnetic waves and transmitted to receivers, this communication approach was used directly with human speech. Now the human voice was transmitted to remote locations, thousands of miles away, picked up by receivers and converted to speech by speakers. Radio was the first complete electronic communications system and it gave a person the capability of "yelling" from one part of earth to another or to "throw" one's voice from a point on earth to some point in space, such as the moon. Further, the travel time for the message was reduced to that available with the speed of light so that international communications occurred within fractions of seconds, and space communications occurred within seconds.

COMMUNICATING WITH LIGHT PATTERNS

The first forms of visual communications occurred through facial expressions and body positions. From this point visual communications split into several alternative techniques. Some used arm and hand positions for direct communication; some used light patterns for signaling over long distances, and some used drawn pictures to point the way to food, shelter, or to tell stories.

Hand and Arm Signals

The most common hand signaling approach was through sign language, one of which is shown in *Figure 1-10.* Initially the signs were limited in number and indicated some concept such as peace, friendship, and so on. Today sign language is still being used, particularly for communication with the deaf, even to the extent of having a separate hand symbol for each alphabetical character. *Figure 1-11* shows the pattern for the letter A. The semaphore code is an extension of sign language in which flags are used to send letters of the alphabet. *Figure 1-12* shows the semaphore code for transmitting the letter A. This method has been used extensively by the Navy for communications between ships at sea.

Figure 1-10. *Indian Sign for Friend.*

Figure 1-11. *Deaf Sign for A.*

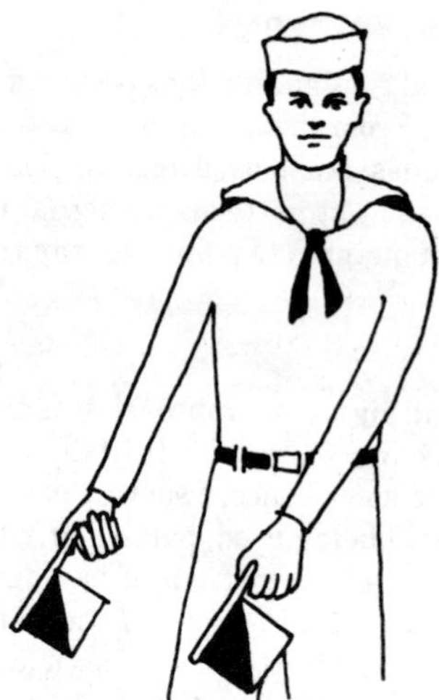

Figure 1-12. *Semaphore Flag Sign for A.*

Codes, Pictures and Signs

The use of light patterns in the form of signs or codes to provide signaling over distances was also used fairly early in man's development. The previously described Indian smoke signals were used for general announcements or warnings. Many early Indian and African tribes used rock or stick patterns to mark trails for others to follow. Even today, hobos or tramps have a trail marking sign language. *Figure 1-13* shows how fences are marked in front of residences to indicate the generosity or receptiveness of the occupant to beggars and tramps.

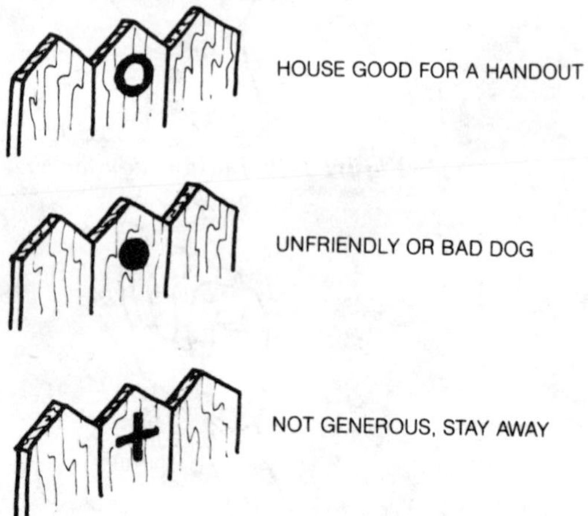

Figure 1-13. *Hobo Sign Language – Fence Signs.*

Light Patterns

Light flashes have been used throughout history to send messages. All of us are familiar with the light signaling used by Paul Revere at the Old North Church in America's revolution. The Navy has used the technique of sending coded messages from ship to ship by interrupting a light beam in a certain pattern. Morse code pulses form the alphabet just as in telegraphy. Transportation organizations such as railroads, airports, and highway and street departments use light patterns for communicating traffic control. The traffic light is one obvious example. Railroad flagmen use lanterns moved in a certain pattern to communicate commands from one end of the train to another – from the flagman to the engineer. *Figure 1-14* shows the patterns used to tell the engineer to proceed and to stop. All of these light signaling methods share the features of low information rate and relatively short range communications. Typically, general commands, warnings, or status reports are being sent. In some cases, alphabetical characters are being sent, but at a fairly slow rate. The light signaling also requires a controllable source of energy in order to achieve the light communication, especially if the range is beyond a hundred feet or so.

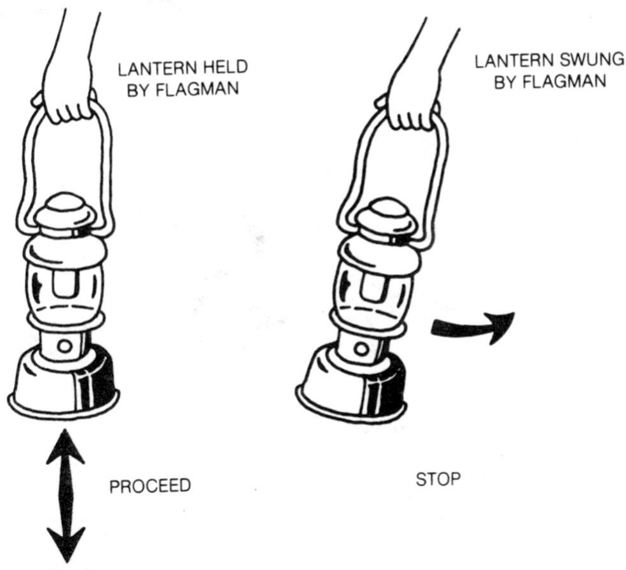

Figure 1-14. *Railroad Flag Patterns.*

Picture Languages and Alphabets

The most advanced form of visual communications have developed from early cave drawings that depicted events in the lives of cave-men. Before 3000 B.C., Egyptians developed a picture language shown in *Figure 1-15* called hieroglyphics and Babylonians developed one called cuneiform, shown in *Figure 1-16*. Actually, cuneiform was a triangular shaped form of writing. Around 2500 B.C. (see *Figure 1-6)*, the Egyptians began using papyrus sheets for writing purposes, which allowed written records of events and governmental activities. The first written alphabet was devised by the Semites about 1000 years later. From that time to the A.D.'s, books and libraries, numerals, shorthand, and writing paper (parchment) evolved. Late in the first 1000 years A.D., the use of ink on paper, block printing, and the present number system came into use, allowing a limited dissemination of the knowledge available at that time.

Figure 1-15. *Hieroglyphics.*

Figure 1-16. *Cuneiform for Mountain.*

Printing Press and Photography

This rate of dissemination of information expanded rapidly once the printing press *(Figure 1-17)* was in operation by the 1500's. By the late 1800's, the development of photography allowed man to communicate through pictures. At about the same time the invention of the typewriter allowed the authors of written communication and information to become fairly efficient in generating the information. All of these developments had a great impact on enabling people to communicate with each other. This was most important in exchanging scientific knowledge. It continued to fuel the ever increasing rate of progress in developing a person's understanding of the physical world. This in turn enabled people to make rapid progress in developing more advanced communications techniques. With these techniques came the capabilities of transmitting pictures and written documents over long distances electronically, first through the use of facsimile (the process of sending printed material or still pictures by telephone for reproduction shown in *Figure 1-18*) and later through the use of television. The further development of satellite communication links shown in *Figure 1-19* provided a means for all types of information to be transmitted from a point on earth to another halfway around the world within fractions of seconds, typically 1/10 to 1/5 of a second. These later visual communications techniques had the advantages of communicating large amounts of information over large distances with small time delays. Such transmissions handle information at millions of bits per second. Compare this to an average of 10 bits per second for information sent by a telegraph operator. It was primarily through the use of electronic techniques that these tremendous improvements in people's communicative ability occurred.

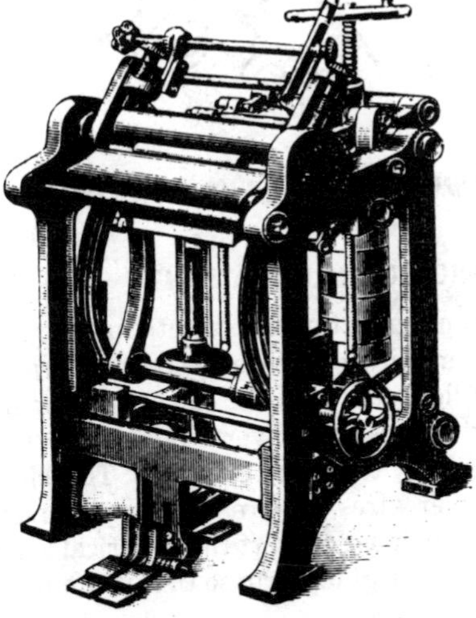

Figure 1-17. *Early Printing Press.*

Figure 1-18. *Facsimile.*
(Courtesy of Xerox® Corporation-Automatic Telecopier®)

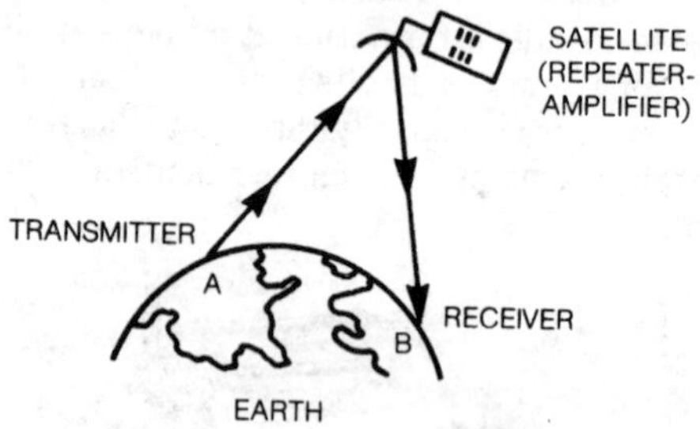

Figure 1-19. *Communication Satellite.*

WHAT IS ELECTRONIC COMMUNICATION?

Electronic communication uses electrical energy to transmit the information to be communicated. Since electrical energy can be made to travel at the speed of light, the transfer can occur within a fraction of a second. The information must be converted from its original form of sound, light, or mechanical energy to electrical energy. This electrical energy is transmitted directly over wires, or wires and amplifiers, or radiated through space to a receiver, which must convert the electrical version of the information back into its original form so that it can be understood by a person or machine.

The basic components or elements of an electronic communication system are shown in *Figure 1-20*. The source of the information can be either a person or a machine. The original form of the information can be a sound pattern such as human speech, a light pattern such as a picture or written document, a mechanical pattern such as a switch closure or a telegraph key or a typewriter keyboard switch, or a physical variable such as temperature or pressure. The original form of information must be converted to the electrical form by an encoder or energy converter. This could be a telephone, a television camera, a switch contact, a pressure gage that converts pressure changes to changes in resistance, etc. Once the information has been converted to electrical form, it may have to be further modified or amplified to a form suitable for transmission over the wire or through space. The receiver must take this form of electrical energy, possibly amplify it to a large enough level, after which it is handled by the decoder or energy converter which converts the information to its final form. This final form, usually the original form of the input to the transmitter, could be the sound in a telephone receiver, the picture on a television set, sounds of short and long durations on a telegrapher's receiver, the printing of information on paper by a typewriter, increasing hydraulic pressure with a pump, etc. The receiver or destination of the information could be a man or a machine.

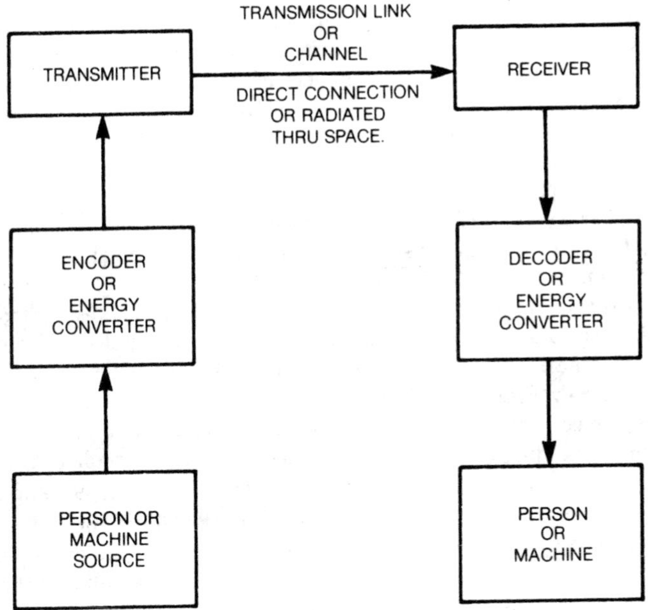

Figure 1-20. *Elements of an Electronic Communication System.*

In the above discussion the possibility of a machine being one or both participants in the communication system is included. This is a real condition since modern communications systems have computers that communicate with each other and people must be able to communicate with computers if they are to use them effectively. Regardless of whether the communicators are persons or machines or both, it is the use of electronic means to provide the information transfer and the features of such a process that make electronic communications so important.

WHY IS ELECTRONIC COMMUNICATIONS IMPORTANT?

The fact that electronic communications is important is relatively evident from the discussion of *Figure 1-5*, but two system characteristics – almost instantaneous information exchange over long distances and versatility – stand out.

Electronic communication has enabled people to interact in a timely fashion on a global level in social, political, economic and scientific areas. This range and immediacy of electronic communications are two of the most obvious reasons why this type of communications is so important. Business transactions, knowledge, political decisions and social reactions can all be communicated rapidly over long range so that a person's intellectual activities are speeded up many times over what they would be without electronic communications means.

An equally important advantage of electronic communication lies in its versatility. Almost any form of information can be encoded into electronic form, transmitted, and then decoded back into the desired final form at the receiver. Speech, music, news, pictures, scientific data, business transactions, military actions, social turmoils, medical telemetry, entertainment, education – all can be handled electronically. As a result, it is almost always possible to establish a communications system that will transfer the exact types of information needed. In fact, the communicator is not limited to a particular system. A choice can be made between several systems to handle the transfer of the information.

It is this range, immediacy and versatility, that makes electronic communications a basic key to success and progress; under the assumption, of course, that nations and individuals use it wisely. For this reason, it is important that everyone learn and understand as much as possible about the basics of electronic communications – not the design details, but enough understanding so that intelligent decisions can be made between the alternatives available so that communications can be efficient and effective. The rest of this book will be devoted to developing such basic understanding.

WHAT HAVE WE LEARNED?

- Communication is the transfer of information in the form of physical patterns from one person or machine to another.
- Much of society's progress in social, economic, and scientific endeavors can be related to improvements in the ability to communicate.
- Electronic communication uses physical patterns of electrical signals to transmit information rapidly and over long ranges from one point to another.
- When evaluating alternative means of communicating, besides basic design parameters, parameters such as rate of information transfer, system reliability, convenience and system cost must be considered.
- Much of society's progress in communicating has occurred since the invention of the printing press and the telegraph and the telephone.
- People communicate by patterns of sight, sound and texture or feel.

WHAT NEXT?

This chapter has dealt with a general introduction to communications and the importance of electronic communications. In the next chapter, the basic means used in electronic communications systems will be explained.

Quiz for Chapter 1

1. Communications between individuals use meaningful patterns of:
   - **a.** Sound
   - **b.** Light
   - **c.** Texture
   - **d.** Any of the above.
   - **e.** a and b above.
2. Encoding of information of communication systems is used for:
   - **a.** conversion of energy to forms that can be used to send the information.
   - **b.** providing privacy in the communications.
   - **c.** increasing the efficiency and reliability of the information transfer.
   - **d.** All of the above.
   - **e.** None of the above.
3. Identify the communications components of a smoke signalling system:

a. Patterns of smoke puffs	**A.** Noise
b. Fire	**B.** Transmitter
c. Blanket	**C.** Receiver
d. Wind	**D.** Code
e. Eyes	**E.** Energy Source
f. Light	**F.** Information Carrier

4. The effect of the ability to communicate on society's progress has been:
   - **a.** nothing important.
   - **b.** important in the areas of social and political structures.
   - **c.** important only in the areas of technological progress.
   - **d.** important in the areas of technology, business and commerce.
   - **e.** b, c, and d above.
   - **f.** b and d above.
5. If a telegraph station receives the following sound pattern in Morse Code, what message was sent?

   ····, · , ·—·· , ·——·

   - **a.** Stop
   - **b.** Help
   - **c.** Love
   - **d.** None of the above.
6. If you are a bum or a hobo and come across the black circular dot marked on a fence post in front of a house, you would:
   - **a.** Stay away from the house.
   - **b.** Try to get work at the house.
   - **c.** Try to get a hand-out.
   - **d.** None of the above.
7. The advent of the vacuum tube and radio communication:
   - **a.** increased the distance of communication.
   - **b.** increased the speed of communication.
   - **c.** enabled one individual to communicate with many individuals over a wide area.
   - **d.** All of the above.
   - **e.** b and c above.
8. Electronic communication techniques:
   - **a.** are limited to special situations in their usefulness.
   - **b.** allow many types of information to be communicated.
   - **c.** are useful only for computer communications.
9. The transmission channel is:
   - **a.** a hole in which the transmitter is located.
   - **b.** the physical path used to transfer the electrical energy containing the information.
   - **c.** a portion of the available electrical energy used to transfer the information.
   - **d.** None of the above.
   - **e.** b and c above.
10. Electronic communication, including telephone and telegraph systems, enabled people to communicate from one continent to another by:
    - **a.** early 1960's.
    - **b.** early 1940's.
    - **c.** early 1900's.
    - **d.** late 1850's.
    - **e.** late 1890's.

(Answers in back of the book)

Using Electrical Signals For Communications

ABOUT THIS CHAPTER

Much of the progress toward today's society has occurred since the introduction of electrical and electronic means of communication. The introduction of radio and television opened the potentials of mass communications and allowed people to transfer large amounts of information over long distances almost instantaneously. The introduction of the computer as a communication system component along with the development of communication systems using satellites located in orbit around the earth led to the establishment of world-wide communication networks. All of these modern electronic communications techniques have had a great impact on the commerce, international relations, science, and entertainment of the world. This chapter will explore the techniques and features of the basic approaches to electronic communication, beginning with an examination of the nature of electrical signals.

HOW CAN ELECTRICAL SIGNALS BE USED TO COMMUNICATE INFORMATION?

In order to communicate, people must generate physical energy or physical properties of matter in patterns that can be detected and understood by another person or a machine. One form of physical energy that has many attractive features is electrical energy. There are several ways that the properties of electrical signals can be varied in order to carry information patterns. To understand the use of electrical signals in communication, let's look at some of the basic properties or features of electrical signals.

Some Basic Electrical Concepts — Current and Voltage

There are two basic types of electrical signals: electrical voltage and electrical current. The relationship between these signals and another basic electrical concept, resistance, is shown in *Figure 2-1*. *Figure 2-1* is called a schematic drawing. It is a way of representing how electrical components are interconnected to form circuits. In the simple electrical circuit of *Figure 2-1* there are two electrical components connected together with wire (the straight lines): a voltage source or battery and a resistor. There are also two measuring instruments shown using a circle symbol: a voltmeter (used to measure voltage) and an ammeter (used to measure current). The battery or voltage supply is the source of electrical energy in this circuit (an electrical circuit is a closed path in which current can flow as a result of voltage applied).

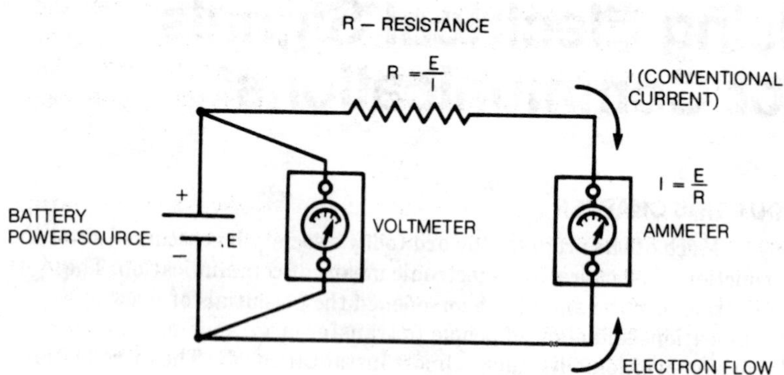

Figure 2-1. *Basic Electrical Circuit Parameters.*

Voltage can be thought of as the pressure in an electrical system. Just as water pressure causes water to flow through a hose or a pipe or air pressure causes air to escape through a leak in a cycle tire, electron pressure causes electrons to flow through a conductor, resistor, or other electrical components. This electronic flow is a flow of charged particles from the negative terminal of the battery (the short line in the battery symbol) to the positive terminal of the battery (the long line in the battery symbol). This flow of charged particles is measured as a current flow using an ammeter, which indicates the current in a unit called the ampere. An ampere is a given rate of flow of charges per second.

While Ben Franklin was studying current flow, before man knew that this current consisted of negatively charged particles called electrons flowing from the negative terminal to the positive terminal of the battery, he guessed at a direction for the flow of the charges. He guessed that positive charges were flowing, and in a direction just opposite to the actual flow. That is, that the charged particles, being positive, flowed from the positive to the negative terminal of the battery. This is still the assumed direction of current today, and it is called conventional current. This current, indicated by the I in *Figure 2-1,* is the current measured by the ammeter in the circuit. This is the convention that is universally used, and of course, it will be used throughout this book.

Some Basic Electrical Concepts — Resistance, Power and Energy

Every element in a circuit, even a wire, offers some resistance to the flow of current. The battery in the circuit must provide enough pressure (voltage) to force the flow of charge (current) through that resistance. The amount of voltage, E, required to force a current of I amps through a resistance of R ohms is simply $E = IR$. This is Ohm's law. If a resistor R is connected across a power supply providing E volts of pressure, then the current I that will flow will be $I = E/R$ amps.

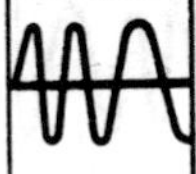

There is another electrical variable that is related to current and voltage and that is power. This is what electrical utility customers pay for when they use electricity. Power is measured in watts, and in circuits that have constant currents and voltages, the power is simply the voltage times the current:

$$P = VI.$$

Many common electrical appliances and components have the power rating listed on them somewhere. Light bulbs are rated by their voltage (usually 110 volts) and their power (such as 60, 75, 100 or 150 watt). Electrical motors are rated in horsepower, with each horsepower equivalent to 746 watts.

Electrical energy is power used over a period of time. Thus, if a 100-watt bulb is turned on for 10 hours, 1000 watt-hours or one kilowatt-hour of energy has been consumed. Electrical utility bills are computed on the number of kilowatt-hours used by the customer each billing period. Power and energy are important variables in communication systems, particularly in the transmitter portion of the system.

The circuit of *Figure 2-1* is of no use as a communication system. Since the voltages and currents are constant, no patterns of voltage or current are available to carry information. The circuit is useful only in terms of defining the basic electrical parameters and their relationships. To provide or carry information, these parameters must change with time in some understandable manner.

Electrical Signals that Vary with Time

The simplest way to make the circuit of *Figure 2-1* useful as a communication system is to provide a way of switching the electrical current on and off as shown in *Figure 2-2a*. By adding a switch and a second resistor to the circuit of *Figure 2-1*, the circuit has been converted to the basic equivalent of a telegraph system. The battery represents the energy source for the system, the switch is the telegraph key, R_1 is the total resistance of the telegraph line that connects the transmitter to the receiver, and R_C is the resistance of the sounding device at the receiving station. Usually the sounding device, a buzzer or relay, requires a certain amount of current for proper operation. The battery must force current through the total circuit resistance which is the line resistance plus the sounding device resistance: $R_1 + R_C$. If I_C amps are required at the sounder, then the battery voltage must be at least $E = I_C (R_1 + R_C)$ volts. Then, when the switch is closed, I_C amps will flow from the battery through R_1, through R_C (causing a click or tone), and back to the battery. When the switch is open, no current will flow, and no sound occurs at the receiver. By closing and opening the switch in a pattern that coincides with a code (the Morse code is the internationally accepted code) for the letters of the words in the message being sent, the sounder will make the corresponding code sounds for decoding or interpreting by an operator at the receiver.

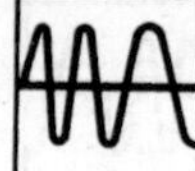

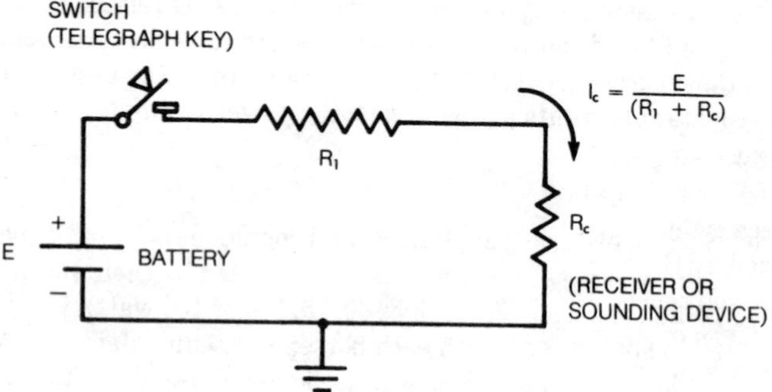

a. Simple Electrical Communications Circuit.

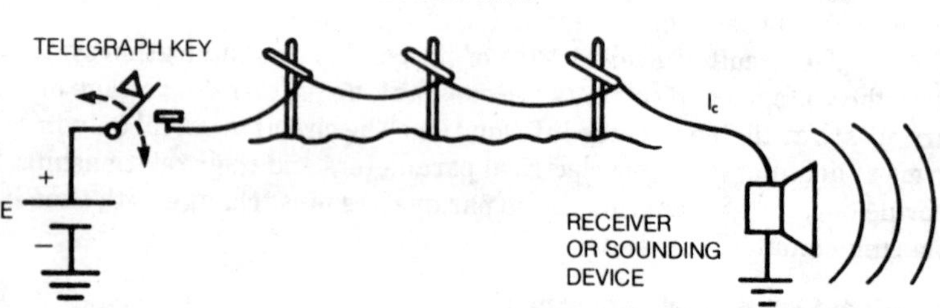

b. Physical Structure Represented by a.

Figure 2-2. *An Electrical Communication System.*

This communication is possible because the circuit of *Figure 2-2b* enables the telegrapher to place a current pattern on the line that varies with time and corresponds to the Morse code of the letters in the message. For example, to send the letter A, which in Morse code is a short pulse or sound followed by a long pulse or sound, the telegrapher would close the switch for a brief period, allow it to spring open, close it for a longer period of time and allow it to spring open again. The resulting voltage variations with time that would be applied to the telegraph line and the resulting current I_C in the line would be as shown in *Figure 2-3*. The electrical line current is carrying the Morse code pattern from the sender to the receiver. Since electrical current travels at about 100,000 miles per second in a telegraph line such as this, the time it takes for the transmission of information from the transmitter to the receiver for a 100-mile line would be only 0.001 seconds or 1 thousandth of a second.

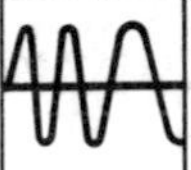

Digital Signals

The types of electrical signals of *Figure 2-3* are called digital signals. They only have two different levels of voltage or current. When the signal is at the first level (usually the most positive voltage) it is said to be ON or TRUE or a 1. when the signal is at the second level, it is said to be OFF or FALSE or a 0. This signal is very commonly used in communications because it is so simple to generate, detect, and use. Since the two levels are fixed by the system that generates them, no information can be implied by the voltage level. The information has to be contained or encoded in the pattern of the levels with time as was done in the Morse code example of *Figure 2-3*.

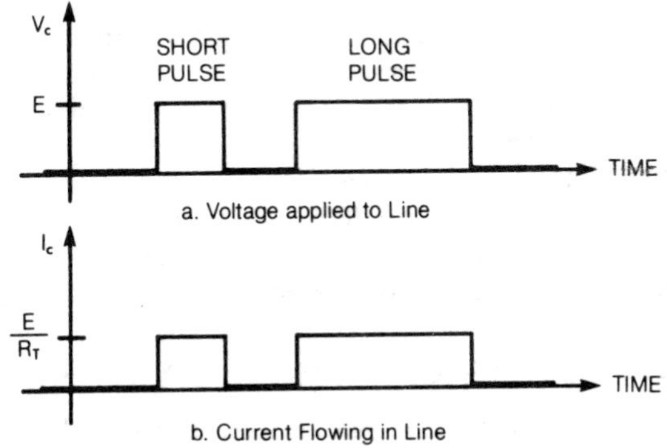

Figure 2-3. *Electrical Signals to Communicate the Morse Code for A.*

The Coding of Digital Signals — PWM OR PDM

There are two basic ways that the patterns of 1's and 0's can be varied to carry information. The Morse code illustrates one method, that of varying the duration of the 1 level. In the case of the Morse code, there are only two pulse widths used, one for a short sound and one for a long sound. This is in itself a digital system since there are only two possibilities to choose from for the information arriving at the receiving end.

This concept can be extended to send more than two levels of information. For example, suppose one desires to send a sequence of decimal digits anyone of which can have a value from 0 through 9. Both sender and receiver decide on the following: a pulse of one second duration represents the digit 0, a pulse of two seconds duration represents the digit 1, and so on up to a 10-second duration pulse representing the digit 9. With this agreement, the sender can generate pulse patterns in order to send numerical information to the receiver. The only thing the receiver must do is time the duration of each sound and write down the corresponding number. Thus, to send the number 132, the I_C would have to be controlled at the transmitter to send the pattern shown in *Figure 2-4*.

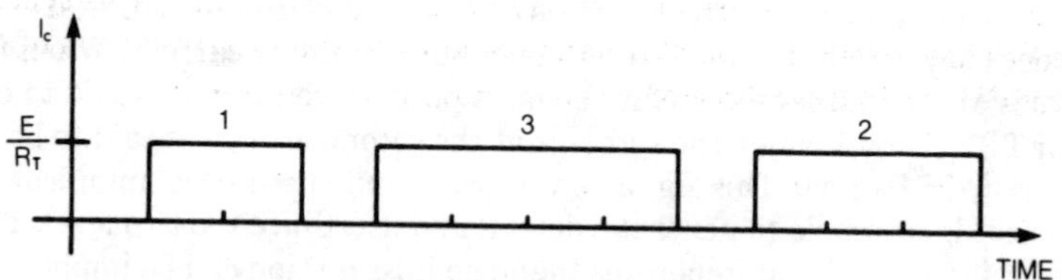

Figure 2-4. *Pulse Width Modulation Pattern to Send Digits 132.*

This is a type of digital coding or modulation of the I_C signal called pulse width modulation. Modulation simply means controlling the amplitude or duration or other characteristics of a signal in a desired way; in this case, causing the "widths" (pulse duration vs. time) of the 1 levels of voltage to behave in the way needed to send the digits 0 through 9 down the telegraph line. Consequently one can see the reason for the name pulse width modulation. Another name for pulse width modulation is pulse duration modulation. These terms are often abbreviated PWM and PDM to avoid having to spell out the entire name of the modulating technique every time.

The Coding of Digital Signals – PCM

A more commonly used method of coding digital signals to carry information using binary (two-level) signals, is called pulse code modulation, often abbreviated PCM. In one version of this technique, the receiver looks at the electrical signal (either voltage or current) on the line at regular periods of time. Previously the receiver and transmitter have agreed on the time when there will be meaningful information present on the line. At that time, there can only be one of two levels on the line, either a 1 or a 0. The receiver records the patterns of 1's and 0's detected at the agreed on times. It must then interpret or decode the meaning of the pattern that has been received.

Look at *Figure 2-5*. In this system, the 1's are all of a fixed duration, centered around the time that the receiver is going to look for a 1. The 1 may be there or it may not be there (not having been sent) at any of the given times at which the receiver measures or samples the line signal. In the example of *Figure 2-5*, the receiver and transmitter agree on a starting time and that the receiver will sample the line every 5 seconds thereafter until told to stop. The transmitter then varies the current I_C, in the pattern shown in *Figure 2-5*, using 1 level signals that are 1 second in duration. Each 5 seconds the receiver would write down whether a 1 or a 0 is received. The receiver would write down the pattern 1001 as a result of the transmission in *Figure 2-5*, look up in a code table (it must be the same as

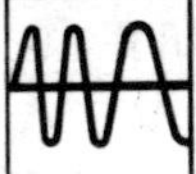

the code table the transmitter used), and interpret the information that was transmitted. In modern day communications equipment of this type, the codes are stored electronically and the look-up and interpretation of the information is all done electronically.

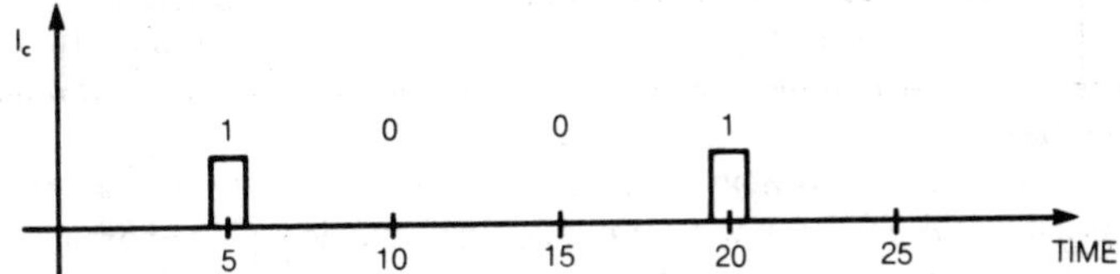

Figure 2-5. *Pulse Code Modulation Pattern to Send 1001.*

One of the most common patterns for the coding of the transmitted information is a binary code. *Figure 2-6a* shows such a 4-bit binary code for decimal numbers. When the receiver detects the pattern of 1's and 0's that it receives in a given period of time (such as the four bits 1001 in *Figure 2-5)* it interprets the number transmitted by finding the corresponding decimal number represented by that code. In the case of *Figure 2-5*, a 9 was transmitted. By changing the pattern of 1's and 0's the transmitter could have sent any decimal number between 0 and 15 with a 4-bit binary code. Alternatively, as shown in *Figure 2-6b*, the sender and receiver could agree that the four bits represent 16 different words or 16 different commands or 16 different letters of the alphabet. The 16 different patterns could represent anything that both the transmitter and receiver agree on.

Binary Code	Decimal Interpretation	Alphabet Interpretation	Command Interpretation
0000	0	K	Stop
0001	1	L	Idle
0010	2	M	Wait
0011	3	N	Step
0100	4	O	Release
0101	5	P	Hold
0110	6	Q	Skip 1
0111	7	R	Skip 2
1000	8	S	Skip 3
1001	9	T	Skip 4
1010	10	U	Alarm
1011	11	V	Time
1100	12	W	Test
1101	13	X	Set
1110	14	Y	Exchange
1111	15	Z	Go

a. Code Table for Decimal Equivalent

b. Alphabet Equivalent or Command Equivalent

Figure 2-6. *4-Bit Code Equivalents.*

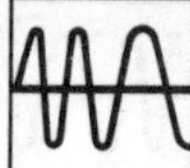

There are longer binary codes and there are different ways in which the receiver and transmitter can agree to sample the signals in order to detect the binary code being sent, but they all have the basic features of the example of *Figure 2-5*, and they are all called pulse code modulation. For now this understanding serves our need, but there are other ways in which the simple digital system of *Figure 2-2* and *2-3* can be used to send information patterns. These are more complicated and will be covered in later chapters.

There are also other ways than the approach in *Figure 2-3* to vary the characteristics of V_C or I_C to carry the information. One approach to be discussed next uses patterns of signal amplitude (either voltage or current) to send information.

Electrical Signal Amplitude Variations and Patterns

In this case, the transmitter has a variety of power supplies that can be switched onto the transmission line in order to present patterns of voltage as a signal on the line. This feature can be provided by modifying the circuit of *Figure 2-2* as shown in *Figure 2-7*. The transmitter has a choice of ten different batteries with different voltage values that can be switched onto the line. If each different voltage represents a digit from 0 through 9, then various numbers or number codes can be sent from the transmitter to the receiver. At the receiver a voltmeter is provided to interpret the digit being transmitted. For example, to send the number 315, the transmitter closes switch 3, releases, closes switch 1, releases, and closes switch 5. Each switch would be closed for a period of time, say one second. This sequence of switch closures would cause the line voltage and line current to take on the waveforms of *Figure 2-8*. Also shown in *Figure 2-8* is the voltage that would be read across the resistor R_C by the voltmeter. This voltage is $I_C R_C$ by Ohm's law. The person receiving the pattern of voltages knows which voltage corresponds to which digit. For example, if R_1 is twice R_C, the voltage across R_C will be one-third of V_1, the line input voltage at the transmitter. Thus with $V_1 = 1$ volt, ⅓ volt at R_C would represent the digit 0; with $V_1 = 2$ volts, ⅔ volt at R_C would represent the digit 1; and so on. The agreement between transmitter and receiver for this information is summarized in table form in *Figure 2-9* just as the pattern of 1's and 0's was for the PCM code in *Figure 2-6*. When the receiver records the sequence of voltages 1.333, 0.667, and 2 volts, by looking at the table of *Figure 2-8*, the information can be interpreted as the original three digits transmitted, 315. This type of encoding information into a signal in terms of voltage or current levels is called amplitude modulation, abbreviated AM.

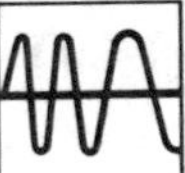

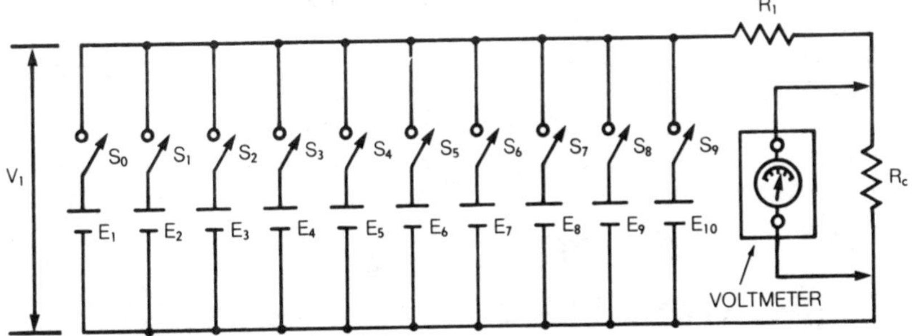

Figure 2-7. Communication System for Sending Pulse Amplitude Signal Patterns.

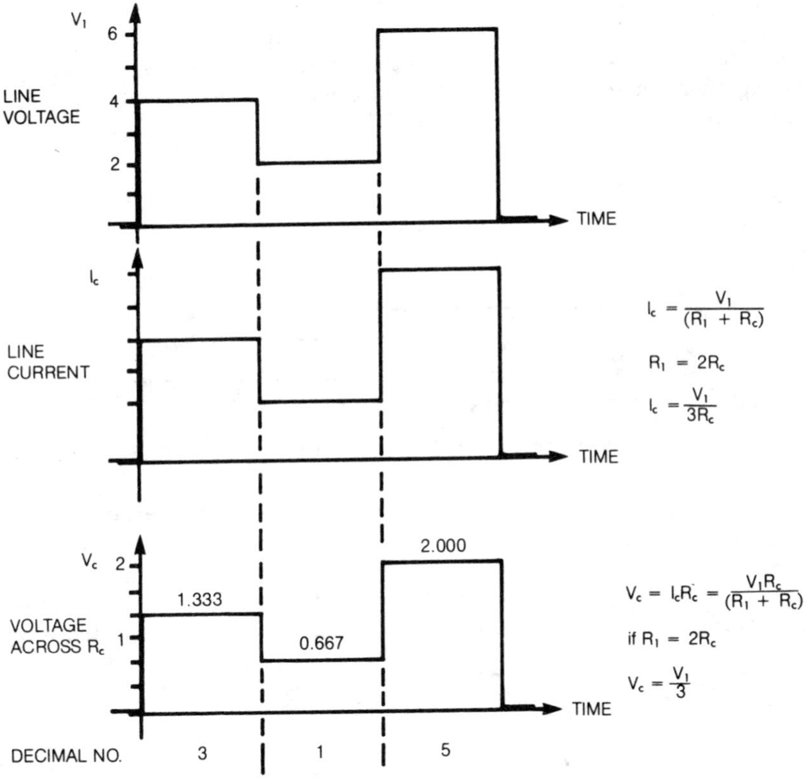

V1
6
4
2
LINE VOLTAGE
TIME

Ic
LINE CURRENT
TIME

$$I_c = \frac{V_1}{(R_1 + R_c)}$$

$$R_1 = 2R_c$$

$$I_c = \frac{V_1}{3R_c}$$

Vc
2
1
VOLTAGE ACROSS Rc
1.333
0.667
2.000
TIME

$$V_c = I_c R_c = \frac{V_1 R_c}{(R_1 + R_c)}$$

if $R_1 = 2R_c$

$$V_c = \frac{V_1}{3}$$

DECIMAL NO. 3 1 5

Figure 2-8. Electrical Signals for Figure 2-7 Needed to Send 315.

V_c Voltage	Decimal Interpretation
0.333	0
0.667	1
1.000	2
1.333	3
1.667	4
2.000	5
2.333	6
2.667	7
3.000	8
3.333	9

Figure 2-9. *Decimal-Voltage Code Table for Communications System of Figure 2-7.*

Amplitude Modulation

Amplitude modulation (AM) can be used to send any type of information, and is not limited to sending just numerical information. In fact, usually in amplitude modulated systems the voltage levels do not change abruptly as in the example of *Figures 2-7* through *2-9*, but they vary continuously over a range of voltage values. One common shape for the way voltages vary in an amplitude modulation communication system is the sinusoid or sine wave shown in *Figure 2-10*.

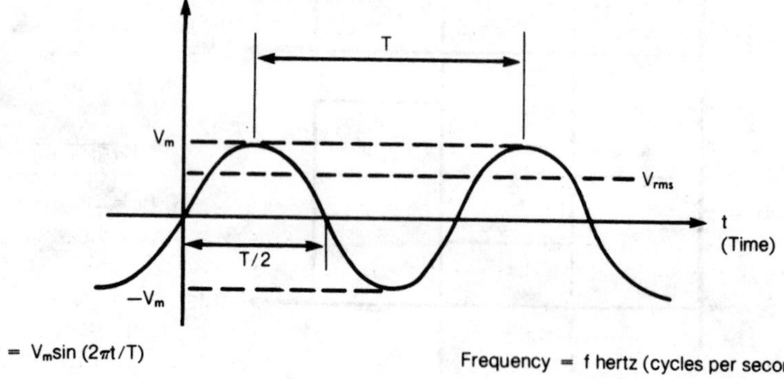

$V = V_m \sin(2\pi t/T)$

V_m = Amplitude in volts

Period = T Seconds

If $t = 0$ $\quad V = V_m \text{ Sin } 0 = 0$

$t = \frac{T}{4}$ $\quad V = V_m \text{ Sin } \frac{\pi}{2} = V_m$

$t = \frac{T}{2}$ $\quad V = V_m \text{ Sin } \pi = 0$

$t = \frac{3T}{4}$ $\quad V = V_m \text{ Sin } \frac{3\pi}{2} = -V_m$

$t = T$ $\quad V = V_m \text{ Sin } 2\pi = 0$

Frequency = f hertz (cycles per second)

$f = 1/T$

$V = V_m \sin(2\pi f t)$

V_{rms} = root-mean-square (RMS) Value

$$V_{rms} = \frac{V_m}{\sqrt{2}}$$

Figure 2-10. *Features of Sinusoidal Electrical Signals.*

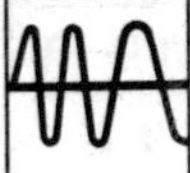

Sinusoidal Electrical Signals

By examining *Figure 2-10,* several of the important features of the sinusoid signal can be defined. The sinusoid varies with time from 0 through a maximum voltage level + V_m, back to zero, through negative maximum of $-V_m$ and back to zero. This pattern repeats with time for as long as the sinusoid signal is maintained. The voltage level V_m is called the amplitude or magnitude of the sinusoid. An important voltage related to V_m and the amount of power that can be supplied by the sinusoid is the R. M. S. (root-mean-square) value, V_{rms}. V_{rms} has a value found by dividing V_m by the square root of 2. The sinusoid of magnitude V_m can deliver the same amount of heat energy to a resistance as a constant voltage of $V_m/\sqrt{2}$.

The sinusoid repeats its pattern every T seconds. This time is called the period of the sinusoid. A related parameter is the frequency of the sinusoid, measured in hertz. This is how many patterns (cycles) occur each second, and thus can be calculated by computing 1 divided by T. Thus, a sinusoid that repeats its pattern once every 1 thousandth of a second (once every millisecond) is said to have a frequency of 1000 hertz or 1 kilohertz. When f is multiplied by 2π (representing the number of angular degrees in radians in a circle) and then by t seconds for $2\pi ft$, the result is an angle in radians. When the time is such that $2\pi ft$ equals $\pi/2$ radians, the sinusoid is at its positive maximum, $+V_m$. When the angle is equal to π radians, the sinusoid returns to zero; when the angle is equal to $3\pi/2$ radians, the sinusoid is at its negative maximum $-V_m$. At 2π radians the sinusoid returns to zero and begins the repetition of the pattern.

Sinusoidal Signals for Digital Data

The sinusoid is a particularly useful electrical signal for transmitting information over long distances. It can be used in many of the same ways that the constant amplitude signals of *Figures 2-3* and *2-7* are used. For example, to send digital signals, the circuit of *Figure 2-7* can be modified. Now instead of E being a battery with a constant voltage level, it is a sinusoidal voltage source of the same frequency but a different amplitude. Only two are used; a large amplitude for a 1 and a small amplitude for a 0. By switching from one voltage source to another in the appropriate pattern, a series of 1's and 0's can be sent. For example, to send the same information as in *Figure 2-5*, the transmitter would switch on the high-amplitude sinusoid for a second centered at five seconds and released. The low-amplitude sinusoid is switched on for a second at 10 and 15 seconds and the high-amplitude sinusoid again for a second centered at the 20-second mark. The resulting pattern of signals of *Figure 2-11* clearly contains the same information as does the pattern of *Figure 2-5*. All that has to be changed in the system are the voltage sources at the transmitter (constant sources to sinusoid sources) and the voltmeter types at the receiving end (constant voltage voltmeters, called dc – direct current –

meters, to sinusoid voltage voltmeters, called ac – for alternating current – meters). The receiver records the magnitude of the received voltage in V_{rms} values to determine whether a 1 or a 0 is sent, when, and for how long. The voltmeter readings when plotted versus time would look like *Figure 2-12*. If the high voltage reading is interpreted as a 1 and the low voltage reading is interpreted as a 0, the original digital information 1001 can be recovered at the receiver.

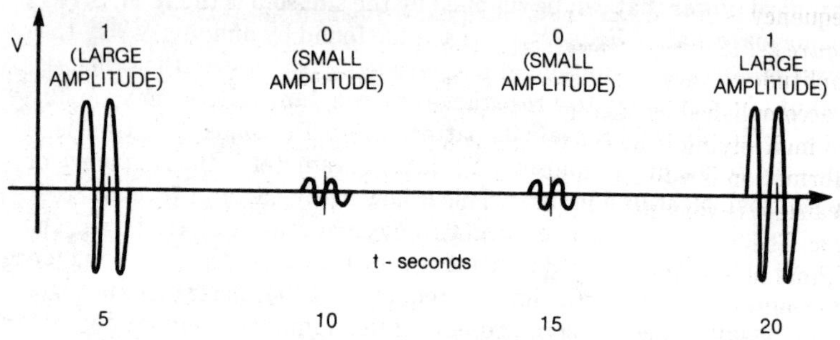

Figure 2-11. *Amplitude Patterns of Sinusoids to Communicate Binary Information.*

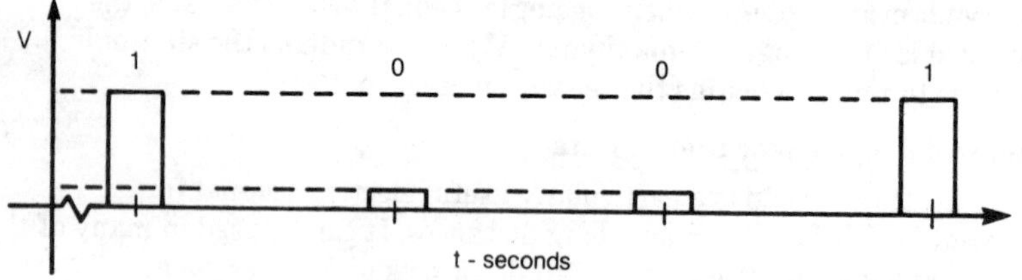

Figure 2-12. *Digital Equivalent of Waveforms of Figure 2-11.*

As before, the system uses amplitude modulation to accomplish the information transmission. It has the advantage over the system of *Figure 2-2* in that the presence of a sinusoid indicates information (a 0 or a 1) is being sent, and the absence of the sinusoid indicates no information is being sent. The receiver does not have to sample the line voltage at prescribed times, even though the example showed it that way. In fact, the waveforms of *Figure 2-11* could be moved together so that the sine wave is continuous, only the amplitude varies. All the receiver need do is monitor the voltage levels and interpret them as 1's or 0's.

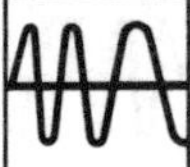

Analog Sinusoidal Signals

Of course, a non-digital signal analog could also be sent using sinusoids. Suppose that the voltage level varying with time of *Figure 2-13a* represents a quantity as an electrical signal that is meaningful in some way to both transmitter and receiver. Because of system performance and ease of transmission and similar factors, it is easier to transmit and receive the information if the amplitude of a sinusoidal waveform of much higher frequency is varied rather than transmitting and receiving the signal of *Figure 2-13a* directly. Therefore, the object is to cause a sinusoid's amplitude to vary in the same way as the pattern of *Figure 2-13a*. This can be accomplished by taking a carrier signal $V_c(t)$ as shown in *Figure 2-13b*, and multiplying it by the information waveform $V_i(t)$ (the (t) means the information is a function of time) of *Figure 2-13a*. This is an analog signal because it is varying continuously with time. The result is shown in *Figure 2-13c*. The amplitude of the final waveform varies in the same way as the original information does. This is transmitted to the receiver. To recover the information at the receiver, a method must be devised to remove one-half of the sinusoid, smooth out the sinusoid and recover the original V_i function. This is a relatively simple task, as the discussion on AM systems in Chapter 4 will show.

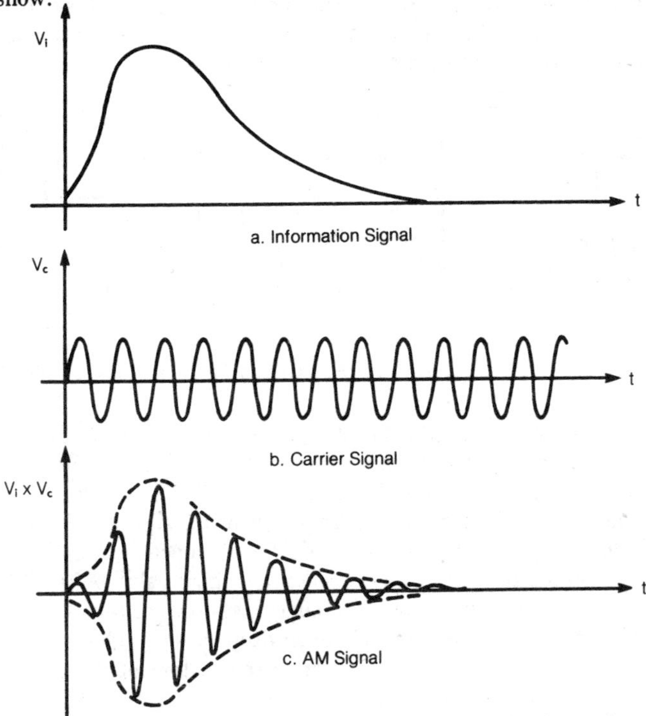

Figure 2-13. *Use of Sinusoidal Carriers to Transmit Analog Information.*

Both of the examples of *Figure 2-13* and *Figure 2-11* involved using patterns of the sinusoid's amplitude to carry information. It is also possible to vary the sinusoid's period or frequency and use these frequency patterns to carry the information.

ELECTRICAL SIGNAL FREQUENCY VARIATIONS AND PATTERNS

The technique now changes. Instead of changing the amplitude of the sinusoidal voltage source, the frequency is changed. The amplitude is held constant. By controlling the frequency, information of the form of *Figure 2-12* or *2-13* is transmitted in the form of patterns of frequency variations. In order for the receiver to be able to detect these patterns, a device that measures frequency must be available at the receiver. For example, if the digital information of *Figure 2-12* is to be sent to the receiver, the transmitter would send an electrical signal of the form of *Figure 2-14*. The high frequency (2Hz) portions of the signal are interpreted at the receiver as a high reading on a frequency meter and are recorded as a 1. The next lowest frequency (1Hz) portions of the signal are interpreted at the receiver as a lower reading on a frequency meter and are recorded as a 0. The lowest frequency (¼Hz) portions of the signal are interpreted by the receiver as having no signal information and are ignored. This strategy by the receiver would result in a reconstruction of the original signal 1001 of the form of *Figure 2-12* at the receiver. Such frequency modulation of digital signals is called frequency-shift-keying (FSK).

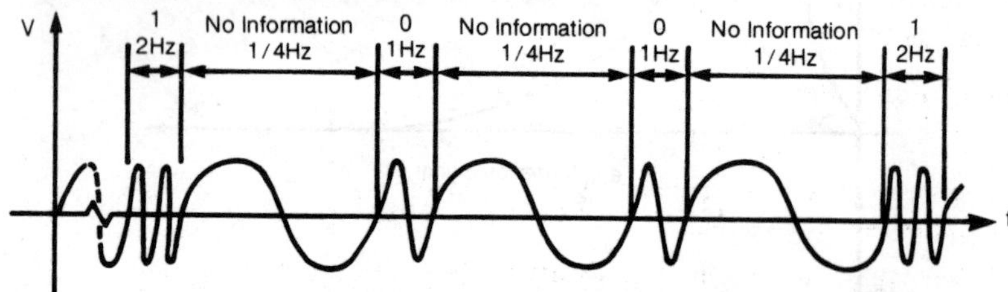

Figure 2-14. *Use of Sinusoid Frequency Patterns to Send 1001.*

The variable amplitude information of *Figure 2-13a* could also be sent using frequency patterns, only the frequency of the sinusoid will be continually varying, as illustrated in *Figure 2-15*. The amplitude of the information V_1 starts at zero, so the frequency of the signal of *Figure 2-15* starts at some low value. As the amplitude rises to a peak value, the transmission signal frequency increases to some high value. After the peak, the amplitude of the information signal decreases toward zero so the frequency of the transmitter sinusoid must drop back to some low value. Electronic means at the receiver reconstruct the pattern of the information signal of *Figure 2-13a* from the rapidly and continually varying frequency in the transmission sinusoid.

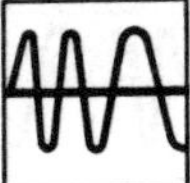

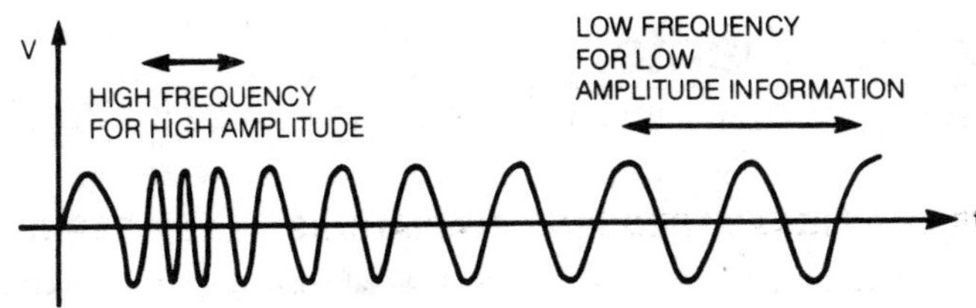

Figure 2-15. *Use of Sinusoid Frequency Variations to Send Information of Figure 2-13a.*

In both the examples of *Figures 2-14* and *2-15*, the technique of sending the information in the form of patterns of sinusoidal frequency variations is called frequency modulation, abbreviated FM. There is a very similar approach called phase modulation, but it will not be covered in detail in this book. Much of what is developed about frequency modulation will be very similar to what happens in phase modulation.

Whether the communication system uses patterns of amplitude or frequency of sinusoids, or whether the basic information is digital (discrete levels) or continuously varying (analog), some means must be provided at both the transmitter to handle the problems of imparting the information patterns to the electrical signals and at the receiver to recover the information from the electrical signals. Fortunately, a wide range of electronic components exist that will handle these problems simply and inexpensively. Most actual communications systems are made from these components.

WHAT HAVE WE LEARNED?

- Electrical signals are of two basic types: analog and digital.
- The basic electrical signal parameters are voltage, current, power, amplitude and frequency.
- Information can be sent electrically in patterns of amplitude or frequency. These patterns can be impressed on the carrier in analog or digital form.
- The sinusoid or sine wave is a basic shape of electrical signals used in communications.

WHAT'S NEXT ?

This chapter has dealt with the basic ways that information can be represented using electrical signals and the basic functions available to manipulate the signals and send information from one point to another.

The next chapter will deal with the components and their functions that are used to carry out the indicated manipulation. It covers the basic functions and the basic types of signal conversion.

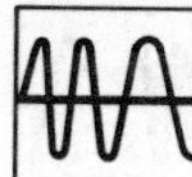

Quiz for Chapter 2

1. In a circuit like Figure 2-1 the battery voltage is 10 volts and the resistance is 100 ohms. How much current flows in amperes?
   a. 1
   b. 0.1
   c. 0.01
   d. 10
2. How much power is delivered by the battery in problem 1?
   a. 1 watt
   b. 1 kilowatt
   c. 0.01 watt
   d. 100 watts
3. In the telegraph system of Figure 2-2 the battery voltage is 100 volts, the line resistance is 90 ohms, and the sounder resistance R_C is 10 ohms. What is the voltage across R_C?
   a. 100 volts
   b. 10 volts
   c. 1 volt
   d. 90 volts
4. What is the frequency of a sinusoid with a time period of 0.001 seconds?
   a. 10 hertz
   b. 100 hertz
   c. 1000 hertz
   d. 0.001 hertz
5. If the transmitter and receiver agree on the command (Figure 2-6) interpretation of a 4-bit code and the code 0111 is received, what command was sent by the transmitter?
   a. Go
   b. Stop
   c. Send Food
   d. None of the above.
6. If an audio signal is to be sent accurately with an 8-bit PCM digital signal, and the audio signal is to be sampled at a frequency of 20,000 hertz, how often (in millionths of seconds or microseconds) should the audio signal be sampled? Once every:
   a. 10
   b. 25
   c. 50
   d. 3.125
7. In the question 6 what would be the bit rate of the digital signal in bits/second?
   a. 320,000
   b. 140,000
   c. 160,000
   d. 200,000
8. With amplitude modulation, the information is placed on a carrier sinusoid by:
   a. adding the modulating information to the carrier sinusoid.
   b. subtracting the modulating information from the carrier sinusoid.
   c. multiplying the modulating information times the carrier sinusoid.
   d. None of the above.
9. With frequency modulation, the information is placed on the carrier sinusoid by:
   a. varying its amplitude.
   b. varying its frequency.
   c. turning the sinusoid on and off.
10. A voltage-to-frequency converter circuit would perform what type of modulation?
    a. amplitude
    b. frequency
    c. phase
    d. PCM

(Answers in back of the book)

Basic System Functions and Conversions

ABOUT THIS CHAPTER

Electrical signals must be transmitted from one point to another. Certain basic functions are required and certain fundamental conversions must occur. This chapter is intended to provide understanding of these concepts so they may be used in specific system descriptions in later chapters.

Many of the basic functions or conversions are available in self-contained packages using integrated circuit technology making it easy to apply them in systems.

WHAT ELECTRONIC COMPONENTS ARE USED IN COMMUNICATION SYSTEMS?

The electronic components that will be discussed are more than single resistors, capacitors, transistors, diodes or other semiconductor discrete parts. They are complete integrated circuits where all the circuitry required for the function is contained on a single chip of silicon and packaged in a plastic or ceramic package. The function performed may be a part of a subsystem, a complete subsystem, or a complete system on a chip. complete subsystem, to a complete system on a chip.

The function of the electronic components for communication systems is to convert electrical signals from one form into another. Here are some examples:

a. Convert steady or dc signals into time-varying signals, either pulses or sinusoids.
b. Convert ac signals such as sinusoids or repetitive pulses into dc signals.
c. Convert analog signals into corresponding digital signals.
d. Convert digital signals into corresponding analog signals.
e. Convert FM signals to voltage levels for information content.
f. Convert information content into FM signals.
g. Convert a low-voltage level signal to a high-voltage level signal.
h. Convert low-power level signal to a high-power level signal.

The last two functions are generally referred to as amplification and occur in almost all systems. Therefore, components providing amplification will be discussed first.

ELECTRONIC AMPLIFIERS

The basic symbol for the electronic amplifier is shown in *Figure 3-1*. Some amplifiers have more features, others have less features than those illustrated by the terminal connections of *Figure 3-1*. The basic amplifier has two inputs: the V_+ terminal is called a non-inverting input because V_{out}, the output voltage, is simply the constant, A, (which is the value of the amplification called amplifier gain) times the signal value on V_+. The V_- terminal is called the inverting imput, because V_{out}, the output voltage, is the negative of A times this input signal (the signal is inverted). As indicated in *Figure 3-1*, when both V_+ and V_- are present, the output voltage is A times the difference between the V_+ and V_- signals.

The value A is a constant in many amplifiers, in which case the V_{AGC} voltage terminal is not provided. When the V_{AGC} voltage control input is available and used, the gain A is a function of this control voltage. This input can be used to stabilize the output signal level at some value or it can be used to perform electronic multiplication as is needed in amplitude modulation. The information signal would be applied to the V_{AGC} terminal in that case.

The amplifier of *Figure 3-1* has a voltage output for a voltage input. There are other configurations possible. Some examples are: amplifier with a current output for a current input, a voltage output for a current input or vice versa. Most of the examples in this book will use the voltage amplifier configuration. The V supply and ground leads allow electrical power to be provided to the circuit. These terminals will be omitted from most of the drawings throughout this book.

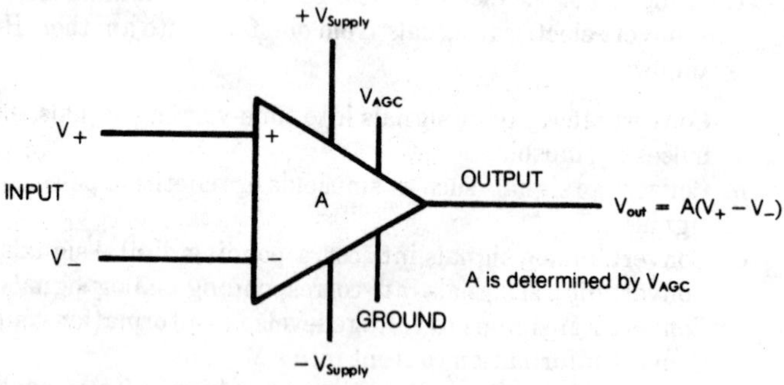

Figure 3-1. *Electronic Amplifier Properties.*

AMPLIFIER FREQUENCY CHARACTERISTIC

Amplifiers usually are classified further by their purpose. Those whose main purpose is to boost signal power from a low to a high level are called power amplifiers. They are classified by the amount of current, voltage or power gain that is provided. The band of sinusoidal signal frequencies the amplifier will amplify classifies it as a bandpass amplifier, an audio amplifier, or a UHF (ultra-high frequency) amplifier. Those that will amplify a wide range of frequencies from very low values to very high values with equal gain are called wide-band amplifiers.

Low-Pass Amplifiers

The band of frequencies amplified may be deliberately limited in one way or another to further classify amplifiers into different categories. Refer to *Figure 3-2a.* If an amplifier is designed to amplify only low-frequency signals and block high-frequency signals, it is called a low-pass amplifier, and its gain-versus-frequency characteristics are as shown. Ideally, these amplifiers should pass all sinusoidal signals below a certain frequency f_c (called the cut-off frequency) by multiplying them by the same constant A. All frequencies above this value should not appear at the output of the amplifier, that is, they would be multiplied by zero. In actual physical amplifiers as shown in *Figure 3-2a,* the gain does not abruptly fall to zero at the cut-off frequency f_c. Instead, the gain gradually decreases from A to zero over a finite range of frequencies; however, very-high frequencies are still blocked and lower frequency sinusoids are still multiplied by A. If the amplifier still amplifies signals with a frequency of zero (which would be a signal with a constant level with time or so-called dc signal), it is called a dc amplifier.

High-Pass Amplifiers

The opposite of the low-pass amplifier is the high-pass amplifier. Its gain-versus-frequency characteristics are as shown in *Figure 3-2b.* With this amplifier, the low-frequency sinusoidal signals are blocked (multiplied by zero) and all the high-frequency sinusoidal signals are multiplied by the same constant A. Again, the difference between the ideal and actual performance of amplifiers is illustrated in *Figure 3-2b.* Most amplifiers would not be able to pass a signal with infinite high frequency as *Figure 3-2b* implies. Therefore, usually the amplifier would have a high-frequency cutoff.

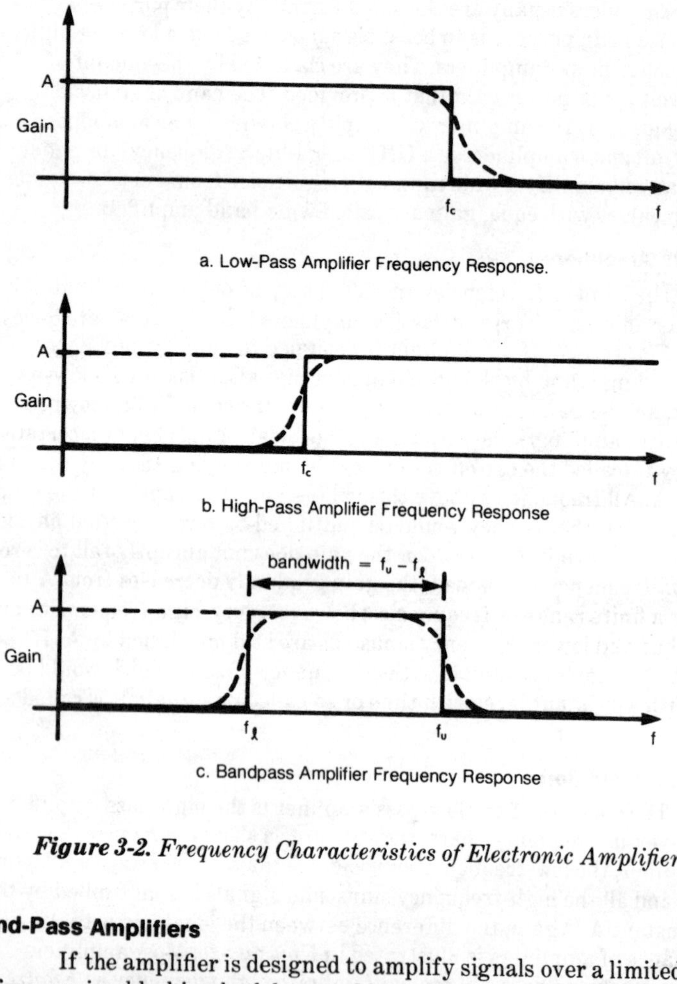

a. Low-Pass Amplifier Frequency Response.

b. High-Pass Amplifier Frequency Response

c. Bandpass Amplifier Frequency Response

Figure 3-2. *Frequency Characteristics of Electronic Amplifiers.*

Band-Pass Amplifiers

If the amplifier is designed to amplify signals over a limited range of frequencies, blocking both low- and high-frequency sinusoids, it is called a band-pass amplifier. Its gain versus frequency characteristics are shown in *Figure 3-2c.* All three types of amplifiers find use in communications circuits, and the reader should become familiar with the features of these amplifiers and the terminology of low-pass, high-pass and band-pass amplifiers.

OPERATIONAL AMPLIFIER CHARACTERISTICS

Summing Amplifiers

Amplifiers can provide other functions that are important in communications systems. They can be connected in a way to perform arithmetic operations such as summation and integration. Those designed for use in such applications are called operational amplifiers. The amplifier of *Figure 3-3a* is connected as an operational amplifier to perform the addition (summing) of three electrical signals. Since the signals are all connected to the inverting input through a resistance of R ohms, the output voltage is the negative of the sum of the input voltages multiplied by a gain factor of R_f/R. Such an amplifier can be indicated with a symbol shown in *Figure 3-3b*. It is easier to use in system diagrams than to draw the actual amplifier connections. It summarizes which signals are to be summed.

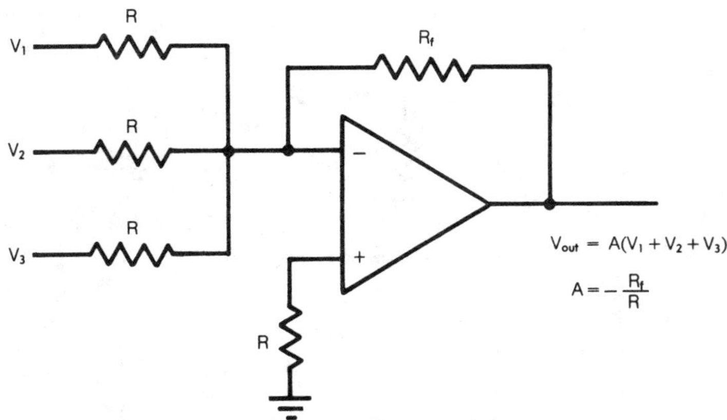

a. Summing Amplifier Connection.

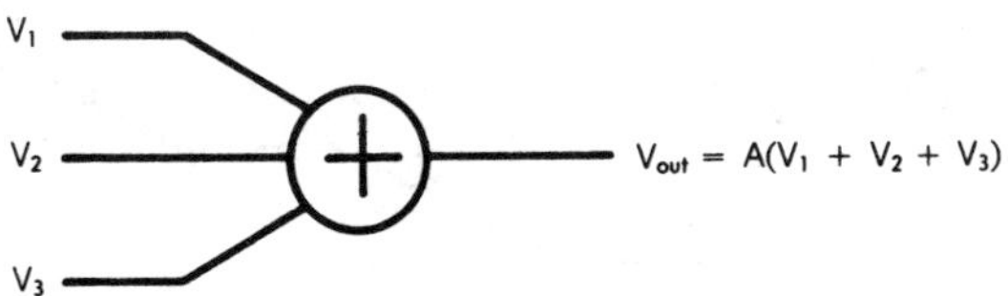

b. Symbol for Summing Amplifier.

Figure 3-3. *Use of Electronic Amplifier to Perform Addition.*

Integrating Amplifiers

Another mathematical operation that occurs in communication systems is that of integration. An amplifier connected as a simple operational amplifier integrator is shown in *Figure 3-4a*. *Figure 3-4b* shows a pulsed input signal V_1 to the integrator and *Figure 3-4c* shows the result of the integration. Integration sums the area under the electrical signal curve. Thus, as each pulse enters the integrator the area under that pulse is added to the accumulated area thus far and the result appears as the output voltage V_{out}. Notice that the effect of integrating a constant, such as the top of the pulses, is to generate a linearly increasing signal. The effect of integrating a signal of zero volts is to leave the output at a constant level.

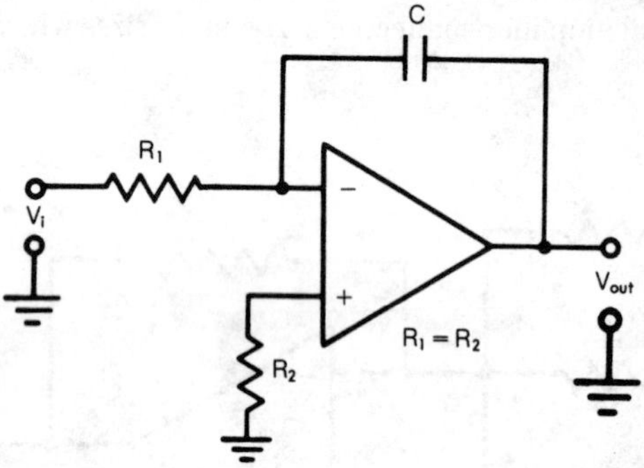

a. Amplifier as an Integrator

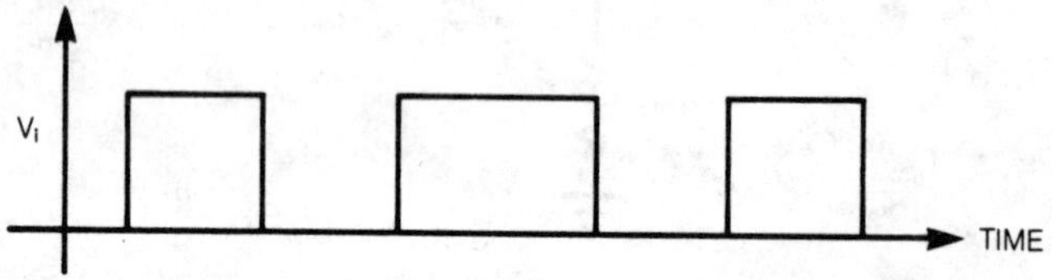

b. Input Signal V_i

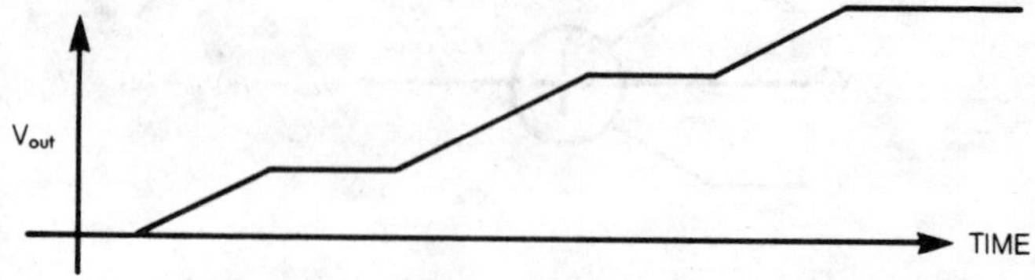

c. Output Signal V_{out}

Figure 3-4. *Result of Integrating an Input Signal V_i.*

Oscillators

One function that is used in all communication systems is that of generating sinusoidal signals of a certain frequency and amplitude. This is called an oscillator. It uses an amplifier in a particular way as shown in *Figure 3-5a*. If the output of an amplifier is fed back to its input so that the input signal is added to or reinforced, a repetitive and periodic electrical signal can be maintained at the output without any additional inputs. In this case the amplifier is acting as an oscillator and is converting power from its dc power supplies into a time-varying, periodic signal. Some of the possible shapes of the time-varying output signals generated by oscillators are illustrated in *Figure 3-5b, c, d,* and *e*. All are repeating patterns with a period T, defined as the time between identical points on the repeating waveform. The square wave (*Figure 3-5d*) is closely related to the sinusoid (sine wave) previously discussed. It is symetrical around zero and has equal positive and negative amplitudes. In fact, all the waveforms of *Figure 3-5* can be constructed from the right combination of sine waves with the proper amplitudes and whose frequencies are some multiple of 1/T.

The square wave and pulse train (*Figure 3-5d, e*) waveforms find extensive use in digital systems. Even the triangular sawtooth waveform, (*Figure 3-5c*), finds application in both analog and digital systems. Of course, these are not the only possible outputs of an oscillator, but they do illustrate the basic properties of the signals that an oscillator may be designed to produce.

SIGNAL CONVERTERS – FREQUENCY/VOLTAGE PATTERNS

Whatever the type of signal produced by an oscillator, a symbol is needed to represent this electronic function in communications systems. *Figure 3-6* shows one possible symbol, with provision for an amplitude control voltage V_A, which is similar to the V_{AGC} voltage in amplifiers, and a frequency control voltage V_f. A sketch of the waveform in the block (some diagrams use a circle) can be used to indicate what type of oscillator is specified. In the case of *Figure 3-6*, the oscillator produces a sine wave output.

Previously it was mentioned that the oscillator is converting dc power to a time varying output voltage. Now with the V_f voltage control of *Figure 3-6* there is another conversion possible – that of converting an input signal voltage into an output signal at a given frequency. Thus, the type of conversion function required to generate an FM (frequency modulated) signal is available. An oscillator that allows a designer to perform the function of converting voltage variations into frequency variations to perform FM is called a voltage-controlled oscillator or a VCO.

The opposite conversion is a function required at a receiver in an FM system, that of converting frequency variations back into voltage level variations. This function is called an FM discriminator in communications jargon.

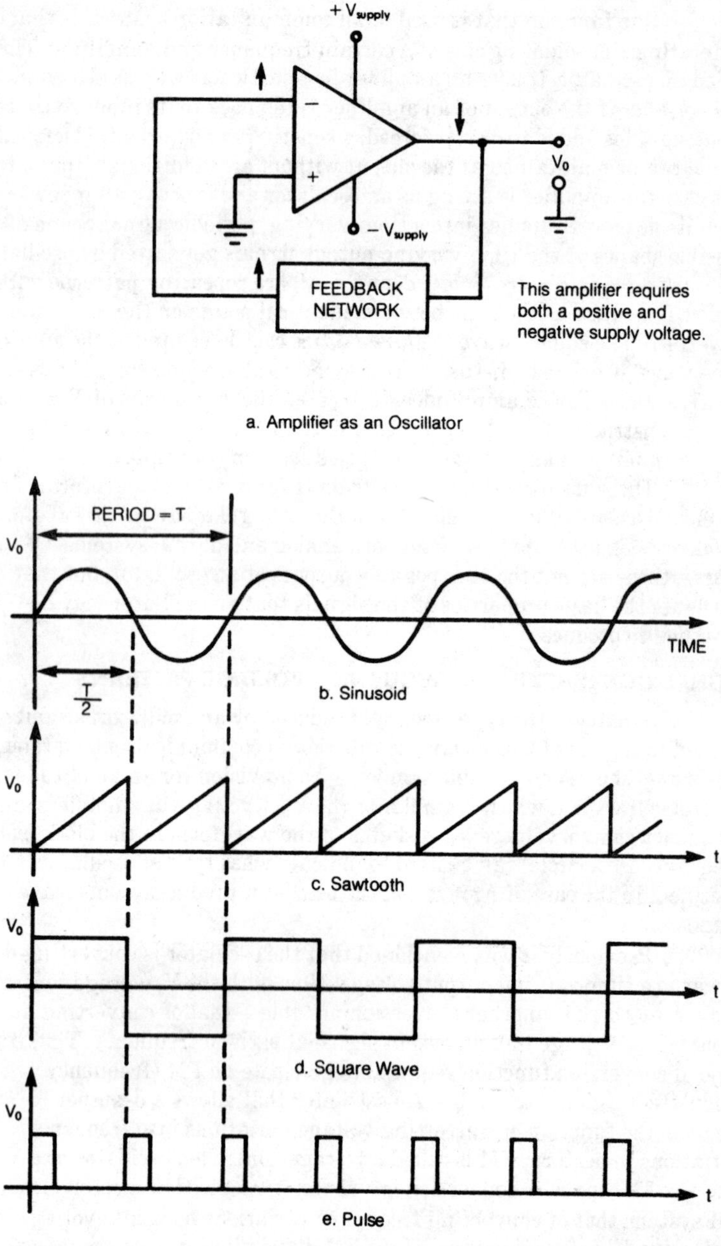

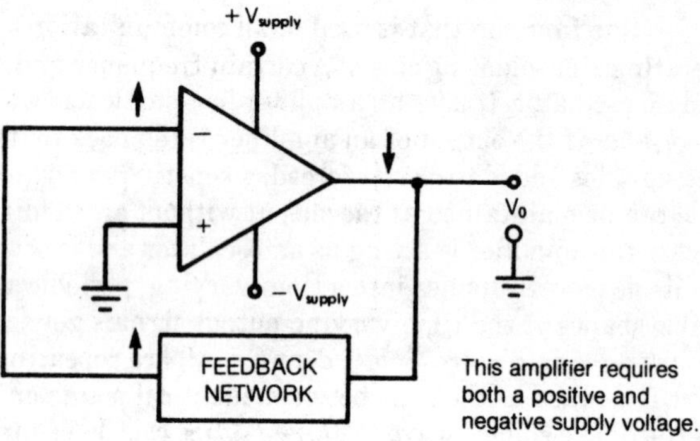

+ V supply
− V supply
V0
FEEDBACK NETWORK
This amplifier requires both a positive and negative supply voltage.
a. Amplifier as an Oscillator
PERIOD = T
V0
TIME
T/2
b. Sinusoid
V0
t
c. Sawtooth
V0
t
d. Square Wave
V0
t
e. Pulse

Figure 3-5. *Possible output Waveforms Produced by Oscillators.*

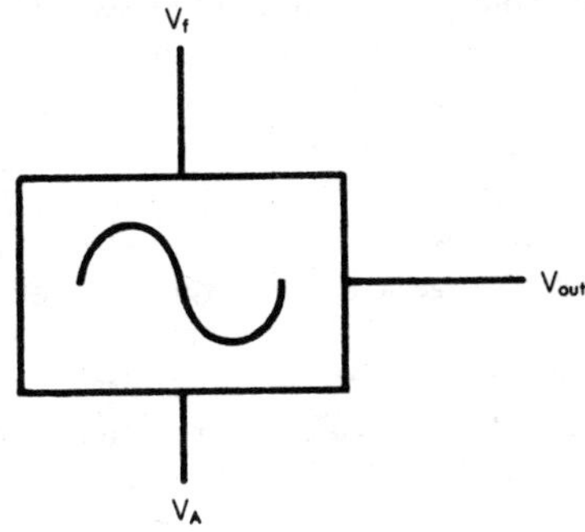

Figure 3-6. *Electronic Oscillator Symbol.*

FM Discriminator

There are several ways such a converter can be built. One subsystem that provides this function is called the phase-lock-loop or PLL. It is shown in *Figure 3-7*. It is a relatively complex combination of some of the electronic functions that have been discussed already. This structure finds many applications in communications systems, one of which is to convert a sine wave signal that has incoming frequency variations to a voltage level variation at one of its outputs. The output voltage variations would correspond to the frequency variations of the input signal.

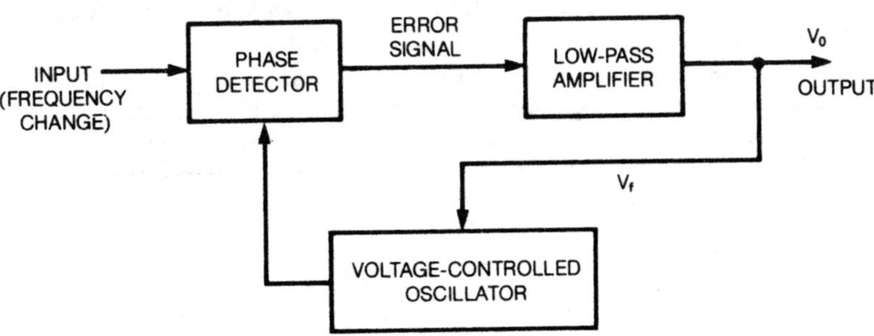

Figure 3-7. *Phase-locked Loop.*

The basic operation of the structure centers around the phase detector and the voltage-controlled oscillator. The phase detector compares the input signal sinusoid with the sine wave generated by the VCO. If the two sinusoids are different in frequency, or if one sinusoid is delayed with respect to the other (a phase difference) an error signal is generated by the phase detector. This causes the voltage V_0 to be changed in such a way as to cause the two sinusoids to match both in frequency and in position in time. When this match occurs, no error signal is generated by the phase detector and the voltage V_0 stops changing. V_0 will then remain at this value until the input sinusoid changes. Each time the input frequency changes V_0 changes to a proportionate value to provide the frequency to voltage conversion.

AM Demodulator

Most signal converters are providing the functions of modulators or demodulators in communications systems. The voltage-controlled oscillator acts as the FM modulator and the PLL acts as the FM demodulator in an FM system. In an AM system the modulator must multiply the modulating signal and the carrier sinusoid. As shown previously in *Figure 3-1,* this can be accomplished by a voltage-controlled gain amplifier in a similar way that V_f controlled the frequency output of the VCO. Now, however, the output amplitude is being varied. The AM demodulation must regain the original modulating signal by rejecting the carrier from the received waveform. Since the carrier frequency is much higher than the modulating frequency in most AM systems, the demodulation is accomplished very simply by using a low-pass amplifier that rejects high frequencies in the band of the carrier frequency and accepts all frequencies in the band of the modulating signal.

SIGNAL CONVERTERS — ANALOG/DIGITAL CONVERSION

One type of signal conversion function needed in communications systems does not totally fit into the modulating and demodulating applications areas. It involves converting signals from analog (continuously varying signals) to digital signals or vice versa. The basic electronic function involved in the conversion function is that of a comparator.

Comparators

The comparator function is illustrated in *Figure 3-8*. If the V_+ signal is greater than or equal to the V_- signal, the comparator outputs a high-voltage signal level or a 1, otherwise the comparator output is a low-voltage 0 level. Refer to *Figure 3-9*. Here the comparator is used to directly implement pulse duration modulation (PDM). A sawtooth sampling signal from an appropriate oscillator is used as the input to the V_- terminal and the modulating information signal is the input to the V_+ terminal. The output is a series of pulses, all of the same height, but of varying widths. Where the modulating signal has a large amplitude, the widths of the pulses are near the maximum. Where the modulating signal is at a low amplitude, the widths of the pulses are near a minimum. Directing the pulse train of *Figure 3-9b* through an appropriately designed integrator at the receiver reconstructs the original modulating waveform.

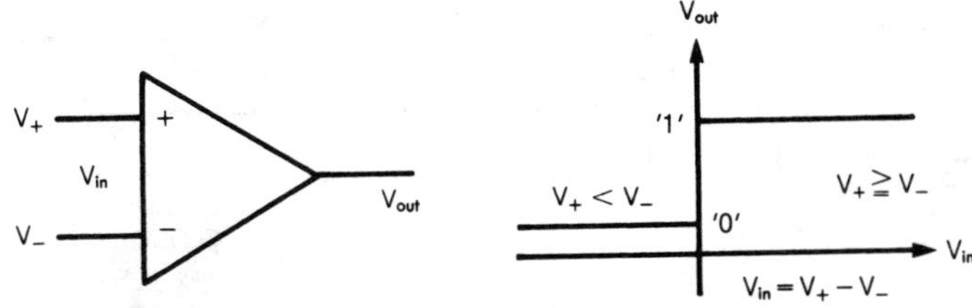

Figure 3-8. *Electronic Comparator Symbol and Function.*

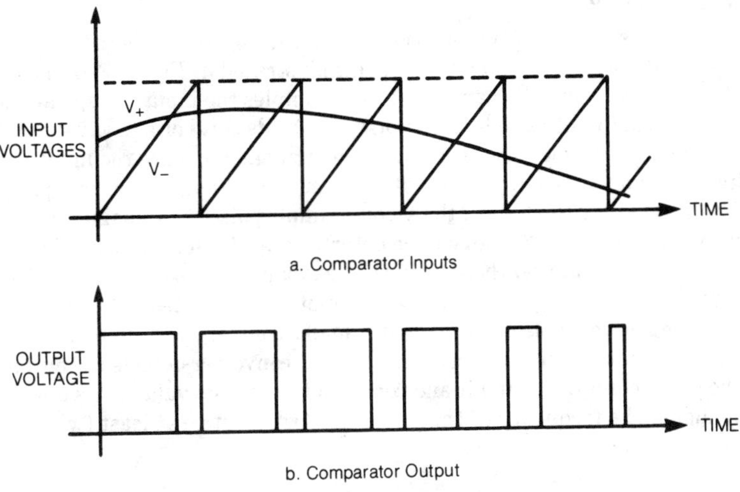

Figure 3-9. *Use of the Comparator as a Pulse Width Modulator.*

The comparator can also be used to generate pulse-code-modulation (PCM) signals directly. *Figure 3-10* is a block diagram of such a system. There are three new electronic functions indicated in this system: a sample-and-hold function, the analog-to-digital conversion function, and a latched register, which is a temporary storage function for digital data. In order to understand the operation of this system, each of these new functions will have to be understood. Note the control input signals called "sample clock" and "latch clock". These are control signals that synchronize the functions of *Figure 3-10*. They cause the functions to occur at a particular time so that operations occur together. As will be apparent from the discussion, they are repetitive pulses occuring at a set time, thus, the name "clock".

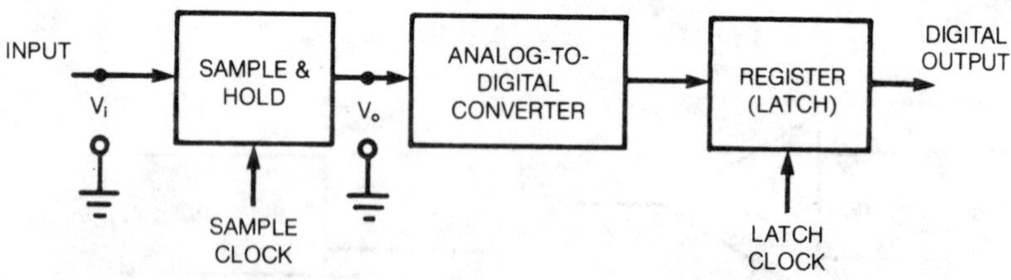

Figure 3-10. *Basic Structure of Pulse Code Modulation System.*

Sample and Hold

The simplest of the functions shown in *Figure 3-10* is that of the sample-and-hold circuit. This operation is illustrated in *Figure 3-11*. The device does just what the name implies: it samples the input voltage at the time specified by the sample clock pulse, and holds this voltage value until the next sample clock pulse. Thus, if at the first sample pulse the input voltage is 7 volts, the output level of the sample-and-hold circuit will jump to 7 volts and stay there until the second sample pulse. At this time the input voltage value of 2 volts will be detected and the output level will jump down to 2 volts, and hold there until the next sample pulse, and so on. The output level will change and hold to the input level detected for each subsequent sample pulse. If the input remains constant for successive samples the output will be constant. The A/D converter detects the output of the sample-and-hold circuit and converts the voltage value to its binary equivalent. The frequency of the sampling clock must be at least twice the highest frequency of the input signal.

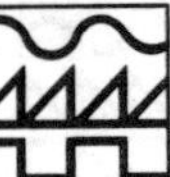

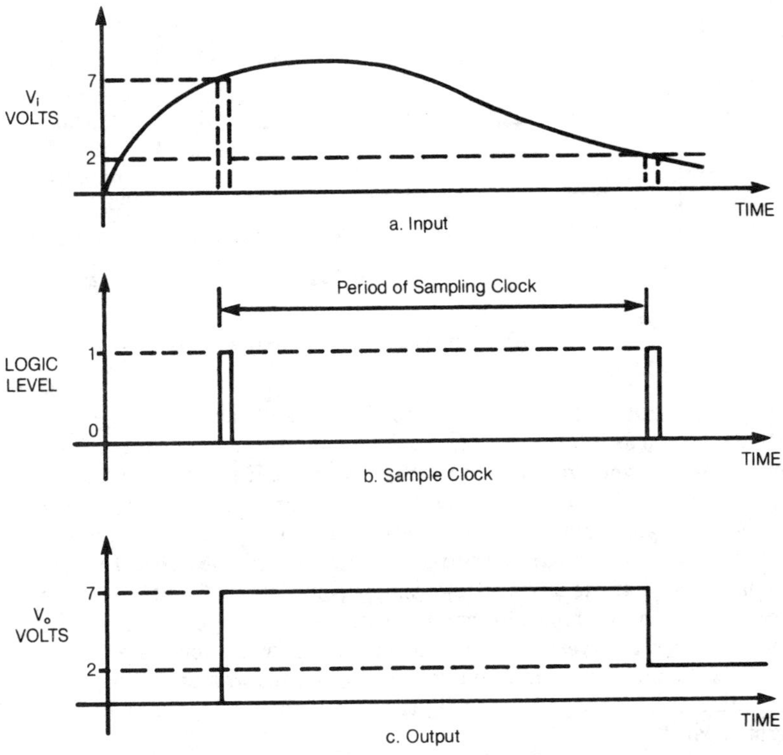

a. Input

b. Sample Clock

c. Output

Figure 3-11. *Operation of Sample & Hold Circuits.*

One method of converting a fixed analog voltage level to its binary equivalent uses a dual comparator device as illustrated in *Figure 3-12.* This is an extension of the application of a single comparator. In this application, two reference voltages are placed on the dual comparator inputs as shown in *Figure 3-12a,* and the signal to be tested is common to the other two inputs. If the test voltage is between the two reference voltages, the dual comparator output level is a 0, otherwise, the dual comparator output level will be a 1. The output versus input relationship is shown in *Figure 3-12b.* It will not be used in this discussion, but for general understanding there is also the possibility of having the comparator outputs reversed (inverted) as shown in *Figure 3-12c.*

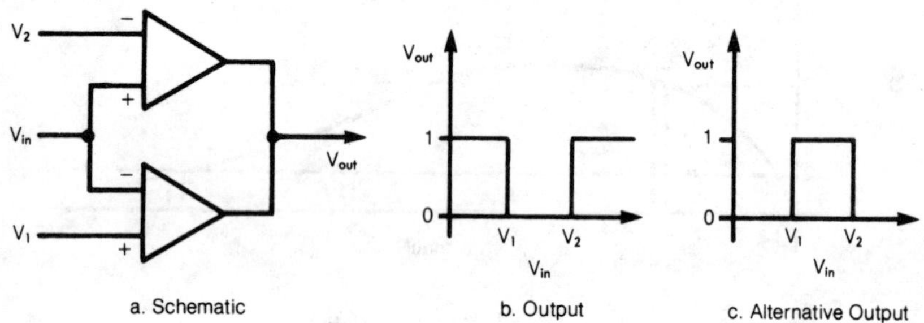

Figure 3-12. *Operation of Dual Comparators.*

If eight dual comparators like *Figures 3-12a* and *b* are interconnected as in *Figure 3-13*, the circuit senses the input voltage and determines the voltage range of the input voltage. If the input voltage is between 0 and 0.5 volts, only the bottom comparator output voltage level will be a 0. If the input voltage is greater than 6.5 volts and less than 7.5 volts, only the top comparator output voltage level will be a zero. In a similar fashion, if the input voltage is between 0.5 and 1.5 volts, the second (from the bottom) comparator will have an output voltage level of a 0. The complete circuit converts an input analog voltage whose level is between 0 and 7.5 volts into its decimal equivalent represented by a digit from 0 through 7. It has categorized (or digitized) the input voltage into one of eight separate levels.

The A/D conversion is not yet complete. In order to process the digital levels with addition digital systems such as computers, it is necessary to convert the decimal numbers into their binary equivalents. To accomplish this requires digital circuits called logic circuits. One such circuit is called a NAND gate. The basic symbol for and summary of operation of the NAND gate are shown in *Figure 3-14*. The table that identifies the output pattern of 1's and 0's for each input pattern of 1's and 0's is called the truth table for the logic circuit, and it completely defines the behaviour of the device. As the NAND gate implies, it is a NOT-AND logic circuit. The output z is not true (a 0) when input x AND input y are true (a 1). Basically, for the NAND gate, if any input goes to a zero, the output goes to a 1; when both inputs are a 1 the output is a 0.

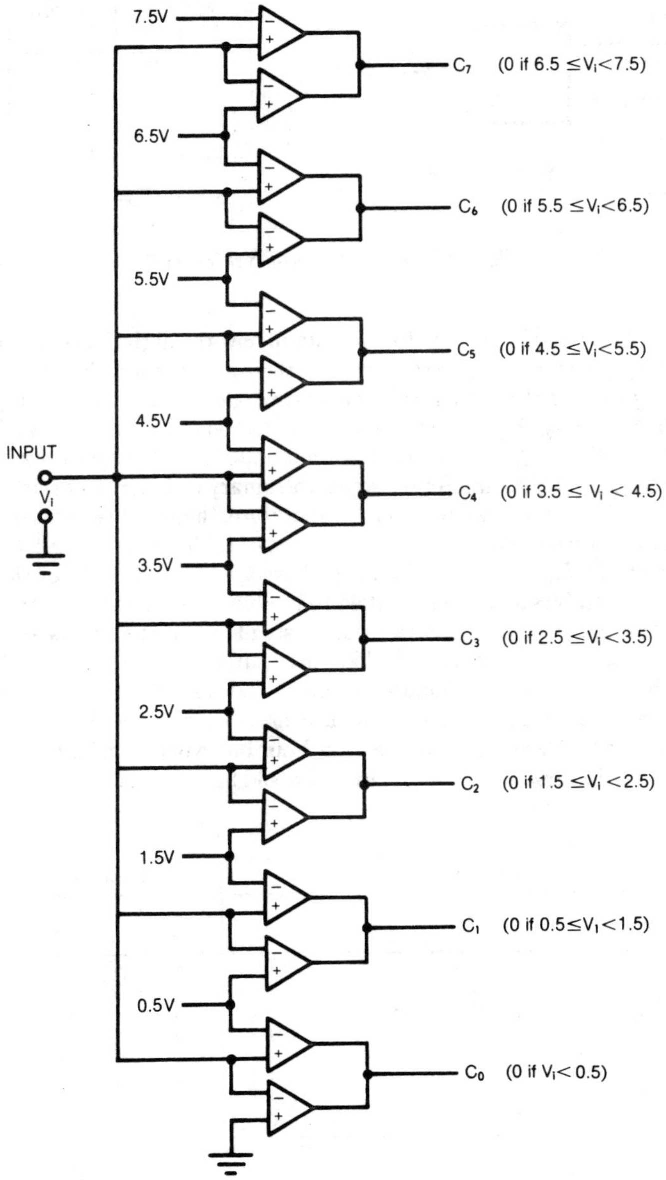

7.5V
C$_7$ (0 if 6.5 ≤ V_i < 7.5)
6.5V
C$_6$ (0 if 5.5 ≤ V_i < 6.5)
5.5V
C$_5$ (0 if 4.5 ≤ V_i < 5.5)
4.5V
INPUT
V_i
C$_4$ (0 if 3.5 ≤ V_i < 4.5)
3.5V
C$_3$ (0 if 2.5 ≤ V_i < 3.5)
2.5V
C$_2$ (0 if 1.5 ≤ V_i < 2.5)
1.5V
C$_1$ (0 if 0.5 ≤ V_i < 1.5)
0.5V
C$_0$ (0 if V_i < 0.5)

Figure 3-13. *Use of Dual Comparators to Digitize Input Voltage to One of Eight Ranges.*

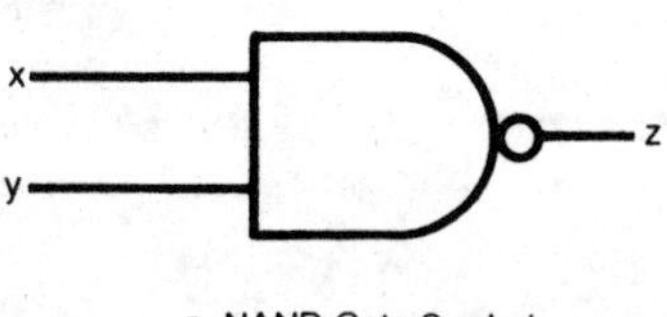

a. NAND Gate Symbol.

INPUTS		OUTPUT
x	y	z
0	0	1
0	1	1
1	0	1
1	1	0

b. NAND Gate Truth Table.

Figure 3-14. *Features of NAND Gate.*

Refer to *Figure 3-15* to understand how the digital level outputs of the A/D converter are to be converted to their binary equivalents. Recall from *Figure 2-6a* the binary code that is equivalent to the decimal digits from 0 through 7. This correspondence is repeated in *Figure 3-15.* V_4, V_2 and V_1 indicate the weighted value of the digit place when a 1 appears in that converter output column. For example, the binary code for the decimal 3 has a 1 in the V_2 digit and V_1 digit places, therefore, the weighted values are added together to obtain the decimal equivalent. The decimal digit 7 has a 1 in each of the V_4, V_2 and V_1 columns because $4 + 2 + 1 = 7$. All of the output column bits must be identified to be able to decode the binary code into its decimal equivalent. As a result, instead of just one line as an output from the A/D converter, there will be three outputs – one for each column. Three bits are required to handle the binary codes for the eight unique digits from 0 through 7. If more unique conditions were required, then more bits would need to be added to the code. Four bits would handle 16 different numbers or letters or commands, 10 bits would handle 1024.

Comparator Outputs								**Converter Outputs**			
C_0	C_1	C_2	C_3	C_4	C_5	C_6	C_7	V_4	V_2	V_1	Decimal Equivalent
0	1	1	1	1	1	1	1	0	0	0	0
1	0	1	1	1	1	1	1	0	0	1	1
1	1	0	1	1	1	1	1	0	1	0	2
1	1	1	0	1	1	1	1	0	1	1	3
1	1	1	1	0	1	1	1	1	0	0	4
1	1	1	1	1	0	1	1	1	0	1	5
1	1	1	1	1	1	0	1	1	1	0	6
1	1	1	1	1	1	1	0	1	1	1	7

Figure 3-15. *Relationships Between Comparator Outputs and D/A Converter Outputs.*

Examining *Figure 3-15* further indicates which output line of the dual comparator circuit is 0 per the discussion of *Figure 3-13*. When the digitized level is 2, the comparator output C_2 has a zero on it; when it is a 5, C_5 has a zero on it; when it is a 7, C_7 has a zero on it, etc.

The table of *Figure 3-15* also shows the correspondence to the code required for the decimal equivalent. Careful examination of the converter output columns shows that whatever electronic circuit is converting the code (pattern of 1's and 0's) on the comparator output to the V_4, V_2, V_1 code must provide 1 in the V_4 column whenever the comparator output code is such that a 0 is on C_4, C_5, C_6 or C_7. Correspondingly, the V_2 column must have a 1 in it whenever C_2, C_3, C_6 or C_7 are at 0; and the V_1 column must have a 1 in it whenever C_1, C_3, C_5 or C_7 are at 0.

Figure 3-16 is an example of how a circuit using NAND logic functions can provide the required binary conversion at the output of the dual comparator. Inputs to a NAND logic circuit (called a gate) providing the V_1 output are connected to outputs C_1, C_3, C_5 and C_7 because, as the truth table of *Figure 3-14b* indicates, a 0 on any NAND gate input makes the output a 1 no matter what the condition of the other inputs. Likewise, the NAND gate inputs for V_2 are connected to C_2, C_3, C_6 and C_7. The overall circuit is called a code conversion circuit because it provides the function of converting from one digital code to another.

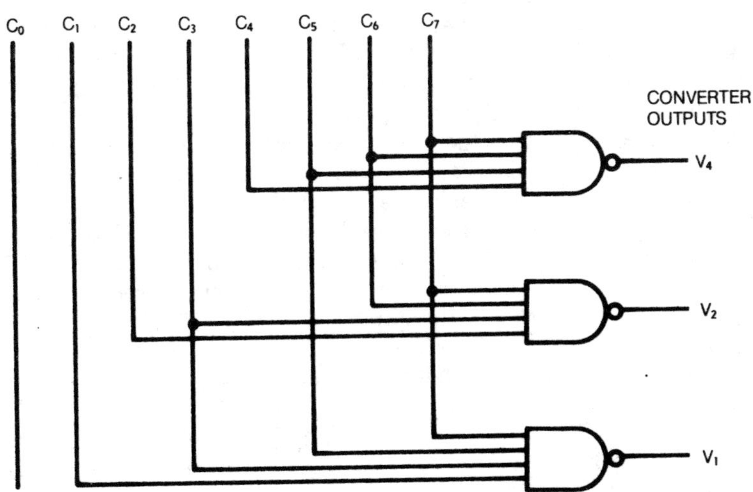

Figure 3-16. *Code Conversion Using NAND Gates*

Latched Register

The outputs of the code conversion circuit of *Figure 3-16* are normally connected to latched registers as shown in *Figure 3-10* so that the binary code may be stored temporarily until ready to be transmitted to the next operation in the digital system. A register is a series of identical circuits placed side by side that are able to store digital information, either a 1 or a 0 as shown in *Figure 3-17a.* A common type circuit that does this is called a "flip-flop" because it is a bistable circuit that flips from one stable state of a 0 to the other stable state of a 1, or vice versa, depending on the state it is in and the state of the input driving it.

In *Figure 3-17a*, input and output gates are shown as part of the register. These are called gates because they allow information to flow when the control lines are active just like a gate controls the flow of water in a channel. With such an input gate, information on the input lines will be stored by the register only when the control line is active. Correspondingly, the information stored in the register will go to the output only when the output gate control line is active. This is best illustrated by examining *Figure 3-17b.*

At time A all conditions are at 0 level. Input BIT 1 and BIT 2 change to 1 at different times but sometime prior to CLOCK 1. None of the register circuits are effected by these changes because CLOCK 1 is 0. When CLOCK 1 is a 1, it gates in the data on BIT 0, BIT 1, BIT 2 so that FF0, FF1, FF2 now hold the input data, 011. This is called "latching the data". Note that even though the data on the input lines changes again after CLOCK, these changes are not stored by the register until the input gate is clocked.

Likewise, the data stored by the register is not placed on the output lines until the output gate is clocked on.

Figure 3-17c shows the type of register latch that is at the output of the A/D converter. This means that the data out of the code conversion will be latched into the register by the latch clock of *Figure 3-17c* and *3-10.* After the latch clock, the register stores the data until it would be transmitted to the outputs by some output gate. This means that the sample and hold circuit can go on to the next sample and the code conversion circuits can change the output code to new data without affecting the data in the register. Normally, the input gates are built into the register to form the latch while the output gates may or may not be included.

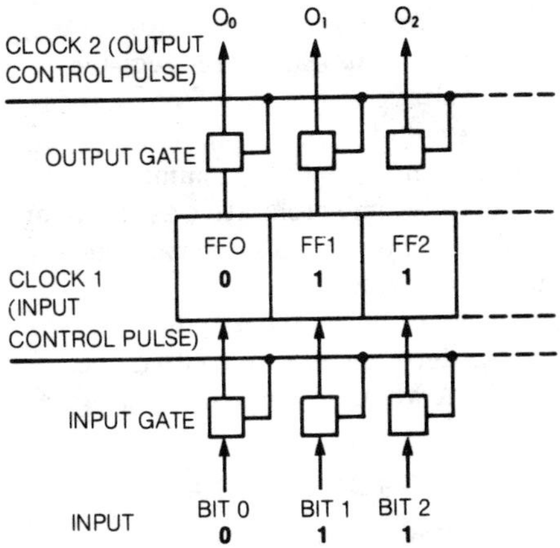

a. Register with Latch Control

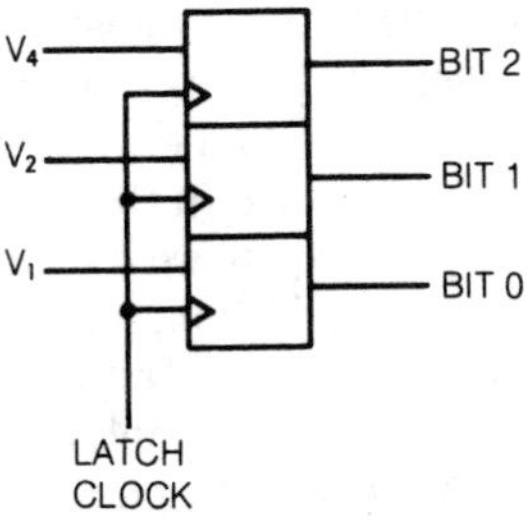

c. A/D Output Register

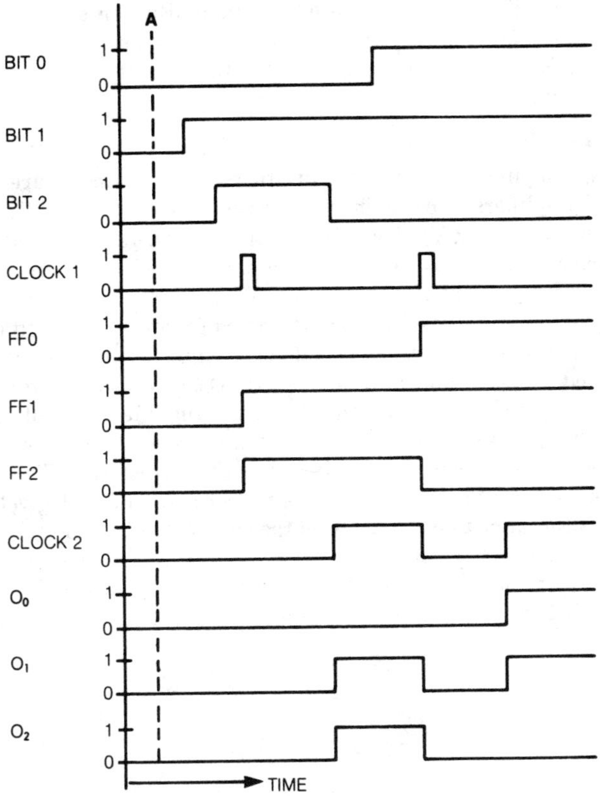

b. Timing Diagram For Latched Register

Figure 3-17. *Register and Latching.*

Figure 3-18 summarizes all of the functions in the overall A/D conversion function. It shows how the 111 code appears at the register output as a result of sampling an input voltage of 7 volts and how a code of 010 appears at the register output as a result of sampling an input voltage of 2 volts. This same approach can be used for an increased number of bits, however, the number of comparators may become prohibitive as the number of bits increases. In this case, more advanced techniques covering the same principles may be used to help keep the hardware simpler.

In addition, some A/D converters have a gated amplifier like the one of *Figure 2-16*. In this case, the V_{AGC} terminal is gated with a clock signal so that the amplifier has a gain of A when the clock signal is present but otherwise has a gain of zero.

SIGNAL CONVERTERS — DIGITAL/ANALOG CONVERSION

After the binary code is transmitted to a receiver it must be converted back to an analog signal to recover the original input. This is another of the desirable functions used in communications systems. Operational amplifiers connected as constant gain and summing amplifiers as shown in *Figure 3-19a* are a major part of the circuitry providing the D/A conversion.

Summing Amplifier

The amplifiers for the input digital code V_4, V_2 and V_1 are operational amplifiers whose gain is set by the ratio of R_f to R_{in}. V_1 is amplified by a gain of 1, V_2 by a gain of 2 and V_4 by a gain of 4. The amplifier outputs are inputs to a summing amplifier with a gain of 1. Therefore, its output will be $4V_4 + 2V_2 + V_1$. As shown in *Figure 3-19b*, such a summation generates the correct analog for any of the 3-bit digital codes that are received and the original analog information has been reconstructed. In many digital-to-analog (D/A) converters the signal levels of the input digital code are such that the input amplifiers are not required for the summing. Special resistor networks or current source networks are usually used instead to combine the inputs in the correct ratio. The output amplifier is used to scale the result to the proper range of analog voltages. The result is the same as obtained with the circuit of *Figure 3-19a*.

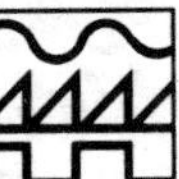

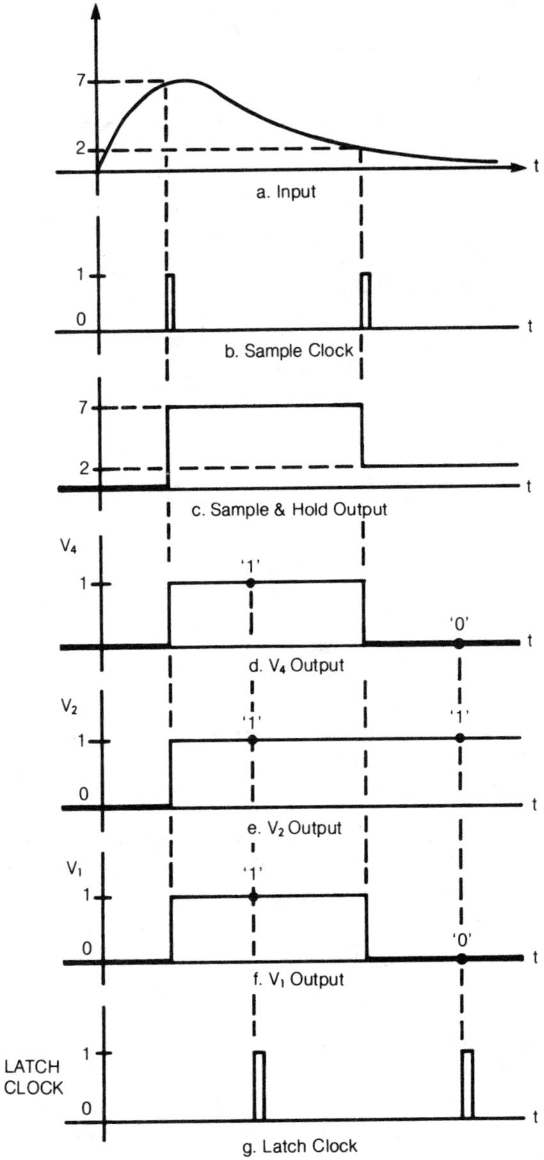

7
2
t
a. Input

1
0
t
b. Sample Clock

7
2
t
c. Sample & Hold Output

V_4
1
'1'
'0'
t
d. V_4 Output

V_2
1
0
'1'
'1'
t
e. V_2 Output

V_1
1
0
'1'
'0'
t
f. V_1 Output

LATCH CLOCK
1
0
t
g. Latch Clock

Figure 3-18. *Operation of A/D Converter of Figure 3-10.*

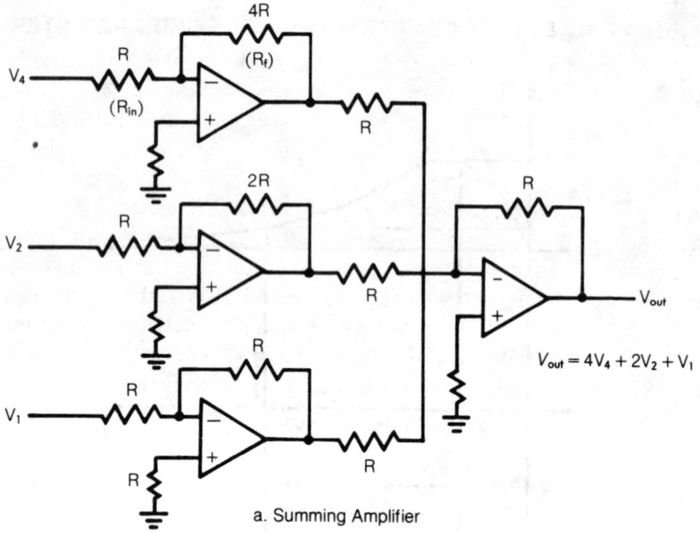

$V_{out} = 4V_4 + 2V_2 + V_1$

a. Summing Amplifier

V_4	V_2	V_1	V_{out}
0	0	0	0
0	0	1	1
0	1	0	2
0	1	1	3
1	0	0	4
1	0	1	5
1	1	0	6
1	1	1	7

b. Conversion Table — Digital to Analog

Figure 3-19. *Basic Digital-to-Analog Converter.*

Serial Data Transfer

One feature of the A/D converter summarized in *Figure 3-18* and the D/A converter of *Figure 3-19* is that the digital information in the form of the 3-bit code flows from output to input, or vice versa, in parallel on a bus of three different wires, as described for *Figure 3-17*. When all the bits of a digital code are sent on separate wires at the same time, the data transfer is said to be a parallel transfer. Most communications systems using wires for information carriers only provide one wire from the transmitter to the receiver. As a result, each bit must be sent on the same wire, one after another, as was done in *Figures 2-11* and *2-12*. This type of transfer is called a serial data transfer, since the bits in the code are sent serially, one after another, on a single wire. Parallel data flow or movement or transfers may be used internally in digital systems, but when an input or output requires serial data transfers, a new function is required that converts parallel digital data to serial digital data and vice versa. Such conversions are called data format conversions.

SERIAL/PARALLEL DATA CONVERSIONS – THE SHIFT REGISTER

The electronic function that can perform the serial-to-parallel and parallel-to-serial data format conversions is the serial-parallel shift register shown in *Figure 3-20.* This shift register is very similar to the register of *Figure 3-17.* The device, in its most general form, provides a parallel data input (like register of *Figure 3-17a*) which is controlled by the parallel load pulse (like input control pulse of *Figure 3-17a*); a parallel data output (like register of *Figure 3-17a*) which is controlled by the output enable pulse (like output control pulse of *Figure 3-17a*); and a serial input and serial output which are controlled with the shift pulse control. It is in this latter provision that the two registers differ, but this provision is what provides the serial-to-parallel or a parallel-to-serial conversions required. As before, the registers provide temporary storage for digital information.

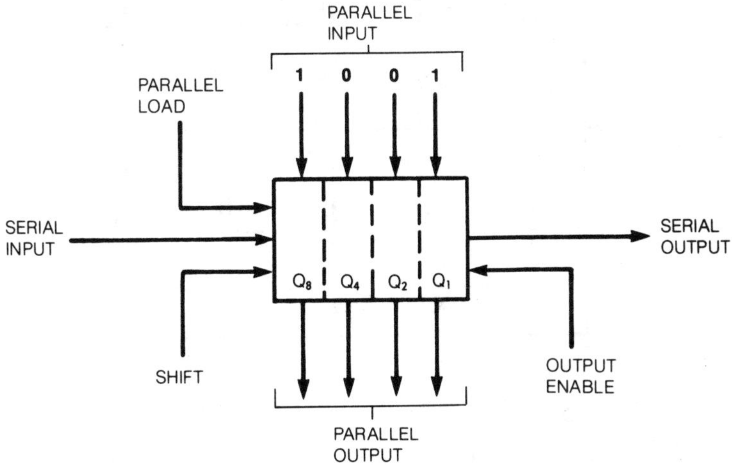

Figure 3-20. *Features of Electronic Shift Registers.*

Parallel-to-Serial Conversion

Figure 3-21 is a timing diagram of voltage waveforms which will be used to describe the operation of the circuit of *Figure 3-20.* First the parallel input data 1001 (*Figure 3-20*), is loaded into the register with the load pulse. Successive shift pulses cause the code to be shifted inside the register, in this case, to the right. As the data is shifted to the Q_1 location it appears at the serial data output. No new data is placed on the serial input; therefore, the data in Q_8 is 0 after the initial 1, loaded by the load pulse, is shifted to the right. Thus, parallel data at the same time (the load pulse time) will be converted to serial data occuring at four separate times (the shift pulse time) on a single output line.

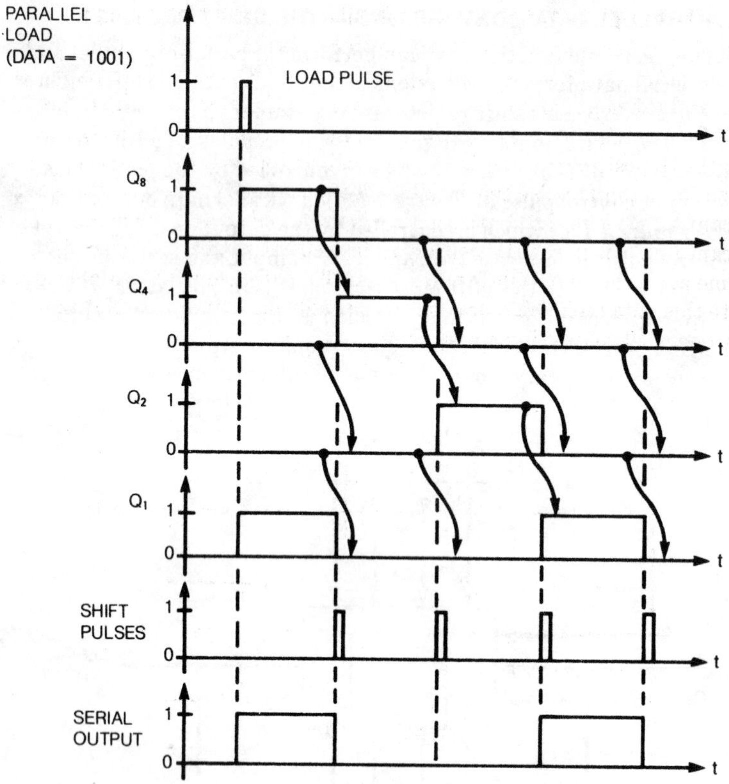

Figure 3-21. *Use of Shift Register to Perform Parallel-to-Serial conversion.*

Starting at the load pulse time in *Figure 3-21*, the least-significant bit Q_1 appears on the serial output line until the first shift pulse occurs, which causes the Q_1 data to become what the Q_2 data had been before the shift pulse. Similarly, the Q_8 data is shifted into the Q_4 position, the original Q_4 data is shifted into the Q_2 position and the original Q_2 data is in the Q_1 position and is appearing on the serial output line. Similar events occur on successive shift pulses so that the code 1001 appears in time on the serial output line as shown in *Figure 3-21* and the conversion is complete.

Serial-to-Parallel Conversion

The reverse conversion, serial-to-parallel, can be accomplished with the same circuitry. This is shown in the timing diagram of *Figure 3-22.* The top line shows the shift pulses as they occur with time. If the data is occurring in serial form on the serial input line in relationship to the shift pulses as shown in *Figure 3-22*, the 4-bit register will fill up with the serial input code 1001 after four shift pulses. This occurs as follows with respect to

time: The least significant bit (1) will be shifted into the Q_8 position with the first shift pulse (it is assumed that all register positions contained 0's before the start). With the second shift pulse, the 1 will move over to the Q_4 position and the next bit (0) will move into the Q_8 position. The third shift pulse will move the least significant bit (1) into the Q_2 position, the Q_8 data (0) into the Q_4 position, and the next bit received (0) into the Q_8 position. With the fourth shift pulse the entire 4-bit input code is received. The least-significant bit (1) is in the Q_1 position a 0 is in Q_2 and Q_4 and the most-significant bit (1) is in the Q_8 position as desired. As shown in *Figure 3-22,* sometime after the fourth shift pulse, a parallel output-enable control signal can gate this data onto the four separate parallel data lines, completing the serial-to-parallel conversion operation.

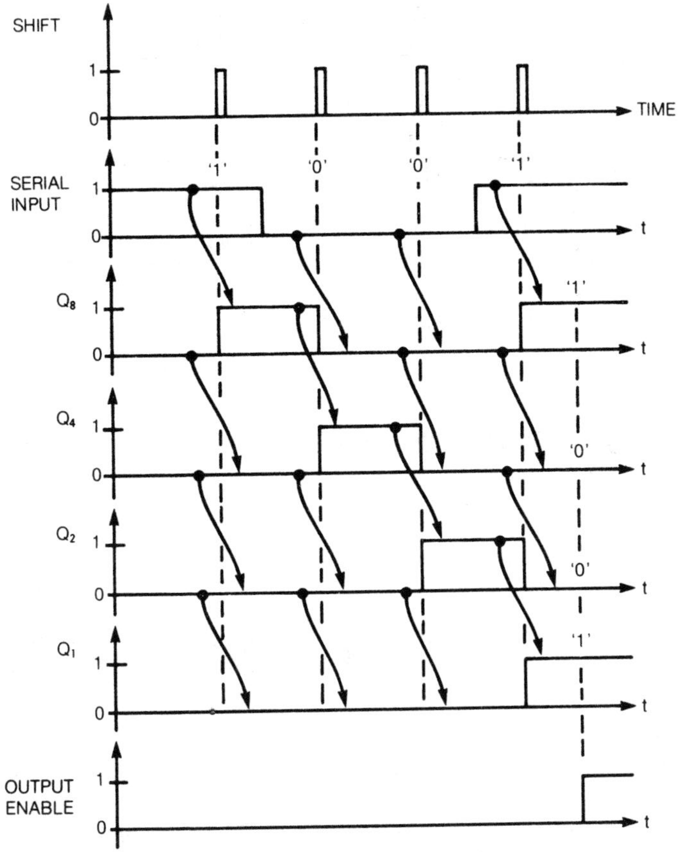

Figure 3-22. *Serial-to-Parallel Conversion Using a Shift Register.*

Combined System

Figure 3-23 summarizes a number of the functions into one system. Analog input information is converted to a parallel 4-bit code by an A/D converter. With a parallel load pulse, the 4-bit code is transferred and stored in shift register A. Shift pulses shift the data onto the transmission line serially and from the transmission line into shift register B serially. After four shifts, the data has been transmitted to shift register B one bit at a time and is temporarily stored in register B. An output-enable pulse sends the 4-bit code in parallel to the D/A converter which reconstructs the original analog input.

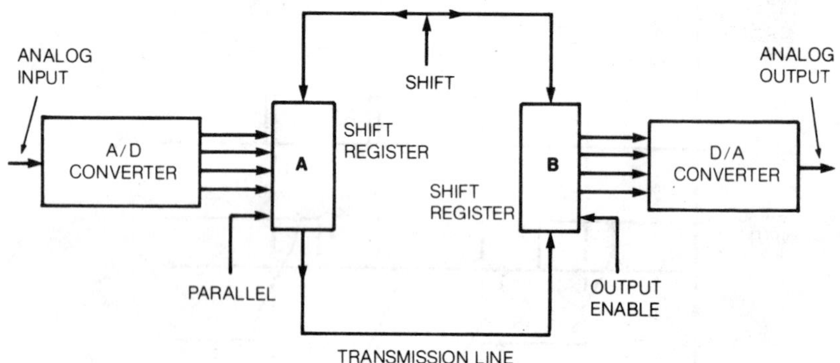

Figure 3-23. *Basic Serial Digital Data Communication System.*

SUMMARY OF ELECTRONIC FUNCTIONS

The basic electronic functions used in communications systems have now been presented. They are not the only ones, but as others are needed in any given communication system, they will be described as they are introduced. These basic functions are readily available and many of them are in integrated circuit form.

As previously stated, with integrated circuit techniques, an entire electronic function is fabricated in a single piece of silicon semiconductor material and packaged in a small lightweight device constructed as shown in *Figure 3-24a, b, c.* Also shown in *Figure 3-24* for comparison is a familiar electrical device, the ordinary light bulb. The light bulb consumes 60 watts of electrical power and provides the simple function of low-level illumination. The electronic integrated circuit consumes only about one-hundredth of a watt (60,000 times less) and performs many complicated electronic operations in a package size reduced by 50 times, with less weight and a reliability that is really not comparable – years instead of months. All this at a cost that is the same or less than the light bulb.

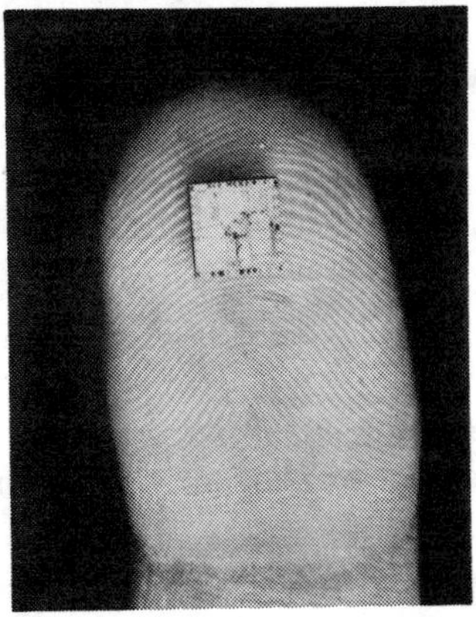

a. Chip

b. Bonded to lead-frame

c. Molded in Plastic Package

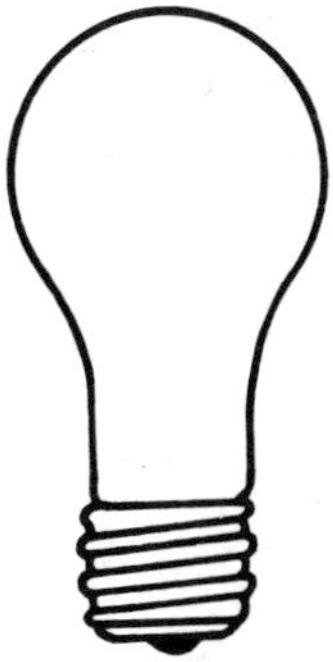

d. Light Bulb

Figure 3-24. *Electronic Integrated Circuit.*

The general purpose of all of the communication systems is to convert information into electrical signals, transmit the electrical signals from one location to a remote location, receive the information at the remote location and recover the information from the electrical signals. Each of these systems uses a variety of the type of functions in different combinations that have been discussed. Using integrated circuits like phase-lock loops, amplifiers, oscillators, and A/D converters, electronic communication systems with sophisticated performance can be produced in small size and at relatively low cost. The system consumes a minimum amount of power and offers the highest reliability or durability.

WHAT HAVE WE LEARNED?

- The basic functions used in communications systems are amplification, oscillation and signal conversion.
- The basic types of signal conversion used in communications systems include steady-state/ time-varying conversions, analog/digital conversions, amplitude/signal conversions, frequency/signal conversions and serial/ parallel digital data conversions.
- The above conversions also have matching inverse conversions.
- There are dc, low-pass, high-pass and band-pass amplifiers.
- A very versatile integrated circuit component providing many functions is the operational amplifier.
- An oscillator is an electronic circuit that converts steady-state dc power to time varying, periodic signals.
- A shift register is used to convert parallel digital data to serial data or vice versa.

WHAT'S NEXT?

Now that there is some understanding of the system functions, the basic communications systems will be examined in a general way in the next chapter. This will help to formulate a mental picture of what is happening to the electrical signals in any one of the systems as they are generated at the source, transmitted, received and converted into the original information at another location.

Quiz for Chapter 3

1. The purpose of electronic amplifiers is to:
   a. increase electrical signal amplitude.
   b. increase electrical signal power.
   c. change signal frequency.
   d. change the signal form from analog to digital.
   e. All of the above.
   f. a and b above.
2. A low pass amplifier rejects signals:
   a. with low frequencies.
   b. with medium frequencies.
   c. with high frequencies.
   d. below a given frequency.
3. A bandpass amplifier is used to:
   a. amplify a wide range of frequencies from 0 to very high frequencies.
   b. amplify a selected range of frequencies and reject others.
   c. pass only musical range frequencies.
   d. pass frequencies down to dc.
4. The AGC control terminal of some amplifiers can be used to:
   a. keep the output signal at a constant amplitude.
   b. perform amplitude modulation.
   c. perform frequency modulation.
   d. keep the signal frequency constant.
   e. a and b above.
   f. c and d above.
5. Oscillators are used to provide:
   a. a periodic waveform of some desired shape.
   b. a sinusoid of a voltage-controlled frequency.
   c. a sinusoid of a fixed frequency.
   d. Any of the above.
6. A PLL can be used as an FM signal demodulator. It contains:
   a. a low pass amplifier.
   b. a voltage-controlled oscillator.
   c. a phase detector.
   d. All of the above.
7. Amplitude demodulation is accomplished with a (an):
   a. PLL.
   b. oscillator.
   c. bandpass amplifier.
   d. high pass amplifier.
   e. low pass amplifier.
8. An A/D converter performs the following operation:
   a. frequency to voltage.
   b. voltage to frequency.
   c. digital to analog.
   d. serial data to parallel data.
   e. analog to digital.
9. A comparator can be used:
   a. as a threshold detector.
   b. as a simple A/D converter.
   c. as a component in a PDM system.
   d. as a component in a multiple-bit PCM system.
   e. All of the above.
   f. None of the above.
10. A PCM system involves the following conversion:
    a. voltage to frequency.
    b. analog to pulse widths.
    c. analog to binary codes.
    d. serial digital to parallel digital data.
11. If the inputs to a quad-input NAND gate are 0, 1, 1, and 0, the output will be:
    a. 0
    b. 1
    c. 1/2
    d. 2
12. Serial data transfers have the following feature(s):
    a. bits occur one at a time on a single wire.
    b. several bits occur at the same time on different wires.
    c. the bits cannot be generated with a shift register.
    d. the bits occur in a matrix in a system memory.

13. Parallel digital data results from:
a. the accumulating of a certain number of serial data bits in a shift register.
b. performing a PCM operation on analog data.
c. doing an analog to digital conversion.
d. All of the above.
e. None of the above.
f. a and c above.

14. If the code 1001 is a digital code that represents a decimal equivalent analog level in volts, the nearest integer to the analog voltage would be:
a. 8
b. 3
c. 9
d. None of the above.

15. An integrated circuit is:
a. an interconnection of components to form a circuit on a printed circuit board.
b. not very reliable and very expensive.
c. a fairly large device or system with integrated functions.
d. a small device with all components and connections provided in a single piece of semiconductor material in a plastic, ceramic, or metallic package with metallic leads for connection.

16. An amplifier that passes all frequencies from dc to a high frequency is called:
a. a bandpass amplifier.
b. a high pass amplifier.
c. a dc amplifier.
d. a wideband amplifier.
e. b, c, and d above.
f. c and d above.

17. An integrator output (Figure 3-4) changes from 2 volts to 3 volts on receipt of 1 volt pulse of 2 microseconds width. What will the output be if it next receives a 2 volt pulse for 6 microseconds?
a. 4
b. 5
c. 6
d. 9

18. An operational amplifier can be used as a(n):
a. oscillator.
b. wideband amplifier.
c. integrator.
d. differentiator.
e. bandpass amplifier.
f. summing amplifier.
g. All of the above.

19. The cut-off frequency of an amplifier is:
a. the voltage above which the amplifier gain decreases to zero.
b. the voltage below which the amplifier gain decreases to zero.
c. the frequency above which the amplifier gain decreases to zero.
d. the frequency below which the amplifier gain decreases to zero.
e. a or b above.
f. c or d above.

20. A digital gate:
a. converts analog voltages to their digital equivalent.
b. determines if a voltage is analog or digital.
c. provides a digital output that is some function of a group of digital inputs.
d. provides a digital output that is a function of analog inputs.
e. provides an analog output that is a function of a group of digital inputs.

(Answers in back of the book)

Basic Electronic Communications Systems

ABOUT THIS CHAPTER

There are three basic types of communications systems: systems that use amplitude modulated carriers; systems that use frequency modulated carriers and systems that use digital techniques. In each of these systems there are many variations. Each of these will be discussed generally to provide an overview to the basic system to be used later as more specific systems are discussed in greater detail.

THE BASIC ELECTRONIC COMMUNICATIONS SYSTEMS

Figure 4-1 shows the general structure of an electronic communications system. At the input there is an energy converter to take information and convert it into electrical signals. The information may be converted to continuously varying analog information, to digital information with binary levels, or to digital information coded with pulse widths or pulse codes.

The energy converter may place the electrical signals directly on the line to an amplifier that provides power boost, or there may be a modulator in the system so that the electrical signals carrying the information are impressed upon some other carrier signal. The telegraph system, where the operator closed a key and caused current to flow in the line circuits for a selected period of time required no modulator. AM broadcast band radio signals require a modulator. The electrical signals representing speech or music or other sounds amplitude modulate a carrier frequency in the broadcast band.

The transmission link can be direct as in early telegraph where all the interconnections are by wires, or through amplified links, as is the case in present day telephone systems, or through electromagnetic radiation – the type link used for all radio wave communications. If there is an electromagnetic link, there particularly needs to be a power boost in the transmitter to the signal levels required for efficient radiation from the transmitter antenna.

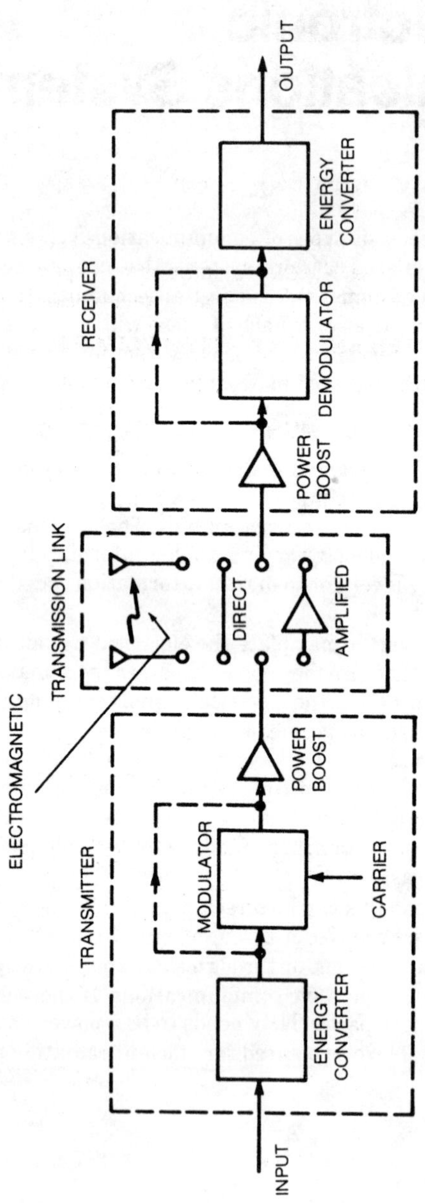

Figure 4-1. *General Structure of an Electronic Communication System.*

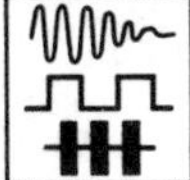

If there is a modulator for a carrier at the transmitter there must be a demodulator at the receiver to provide the reverse conversion. In addition, and possibly more so than the transmitter, power boost is required to restore the weak signals at the receiver to acceptable levels for the energy converter at the output of the receiver. If there is no demodulator, the signal will feed directly from the power boost to the energy converter. If a microphone is used as an input energy converter in an audio system to convert speech into electrical signals, a speaker is used as the energy converter at the receiver output to convert the electrical signals back to sounds – speech in particular. Further examination of specific systems will identify further the functions in the general system block diagram.

AMPLITUDE MODULATED ANALOG SYSTEMS (AM RADIO)

The components of an example amplitude modulated analog system are shown in *Figure 4-2* along with the types of electrical signals at each point in the system. This particular system is communicating sound information such as speech from one point to another. It is the basic system used for AM radio communications.

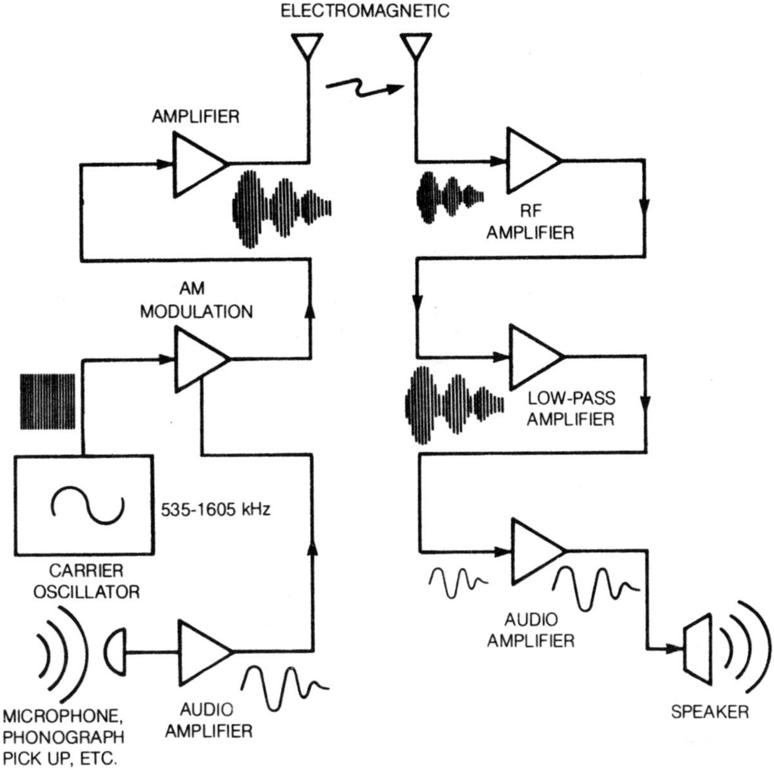

Figure 4-2. *Amplitude Modulation Communication System.*

The Transmitter

The energy converter in the transmitter is the microphone (or tape player or phonograph, etc.). Since the electrical output of such a device is usually relatively low voltage, it must be amplified before it is sent to the modulator. There must be an oscillator that generates the carrier sine wave. For a particular station its frequency is within the AM band from 535 to 1605 kilohertz, many times that of the audio sound frequencies which, over the widest range, extend from 20 hertz to 20,000 hertz. This carrier signal also is sent to the modulator. The modulation process, illustrated in *Figure 2-13*, requires multiplication of the carrier and the modulating signal. This is accomplished with an amplifier like the one in *Figure 3-1* whose gain is controlled by the V_{AGC} voltage. The carrier frequency is the input and the audio frequency is applied to the V_{AGC} terminal. The output of the amplifier is a modulated carrier as shown. A more detailed waveform was shown in *Figure 2-13c.*

Because the transmission link is electromagnetic, the modulator output is boosted to a high power level by the power amplifier and sent to the antenna of the transmitter. This power signal on the antenna is converted to electromagnetic radiation by the antenna. The electromagnetic radiation travels in all directions through space at the speed of light.

The Receiver

At some location, the receiver antenna converts the electromagnetic radiation back to electrical signals. The RF amplifier in the receiver boosts this signal to the level required by the demodulator. In AM systems the demodulator is simply a low-pass amplifier whose cut-off frequency (see *Figure 3-2*) is much below the carrier frequency (minimum of 535 kilohertz) and above the highest frequency (20kHz) in the sound patterns. Thus, the output of the low-pass amplifier contains only the original audio modulating information, since the carrier frequency is not amplified by the amplifier. This audio frequency electrical signal is boosted by the audio amplifier to the power levels required by the speaker. The speaker then recreates the sound that was sent by the transmitter.

Advantages and Disadvantages

One of the advantages of the AM system is its relative simplicity, involving mainly just amplifiers of one type or another and an oscillator to generate the carrier frequency. The only drawbacks to the AM system are its inefficiency in utilizing system power and its susceptibility to noise in the system. Of the total power used in the transmission of an AM signal, 5/6 of the power is not vital to carrying the information. As a result, the transmitter power in an AM system may have to be several times larger than alternative systems to achieve the same range of communication.

Noise

Any unwanted signal not present in the original information can be considered as noise. Noise occurs in all parts of the communication system, and the AM signal is particularly susceptible to noise effects in the transmission link between the transmitter and receiver. Since any signal can be made from a combination of sine waves with the correct amplitudes and required frequencies, noise can be considered to be unwanted sine waves of all frequencies with constant amplitudes that change the transmitted waveform from the original modulated waveform. If the amplitude of the noise becomes comparable to the amplititude of the modulated carrier, the resulting sum of the two looks more like the noise than it does the modulated carrier. As a result is is not possible to reliably reconstruct the original sound pattern from the received signal. FM systems are less susceptible to such noise effects. In addition, the noise problem can be minimized if digital signals are used instead of analog signals.

FREQUENCY MODULATED ANALOG SYSTEMS (FM RADIO)

The basic FM system shown in *Figure 4-3* is comparable to the AM system of *Figure 4-2*. The main change is in the modulator and demodulator devices. In FM systems the modulator must convert the electrical signal containing the information into frequency variations (See *Figure 2-15*). A voltage-controlled oscillator or VCO (*Figure 3-7*) is required for this.

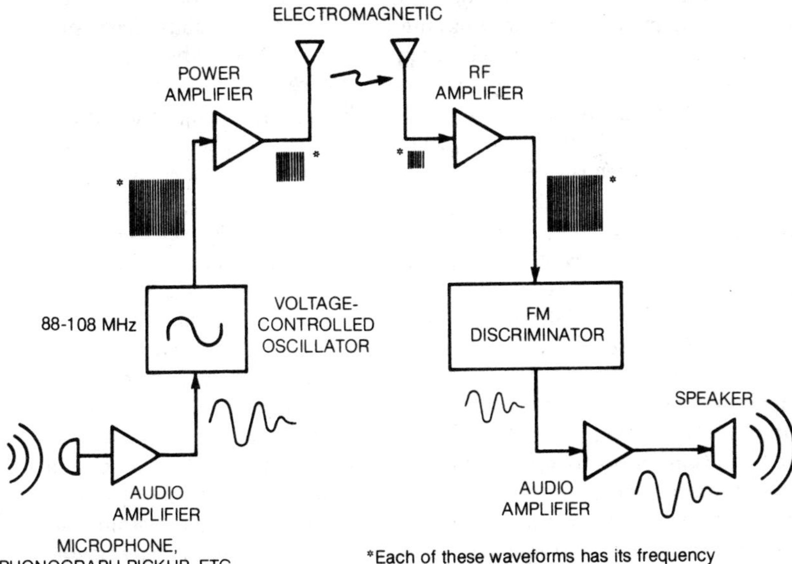

Figure 4-3. *Frequency Modulation Communication System.*

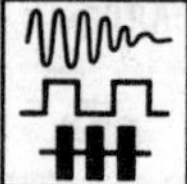

The Transmitter

Just like the AM system, an amplifier is required to boost the signal coming from the microphone or phonograph to the level that is required to operate the VCO. Also, a power amplifier is required to boost the modulator output signals to the level that will generate electromagnetic radiation from the antenna of sufficient strength to cause a detectable signal at the receiver.

The Receiver

At the receiver an RF amplifier is required to boost the very small antenna signals to the voltage levels required by the demodulator which, in this case, is a discriminator such as the phase-lock loop circuit described in *Figure 3-7.* This device generates the same electrical signal that was used to modulate the VCO. When it is sufficiently amplified and sent through a speaker, the original sound waves will be recreated at the output of the receiver.

Advantages and Disadvantages

The FM approach has several advantages over the AM system. The only possible disadvantage has been the increased complexity of the FM modulation and demodulation functions and the circuits to accomplish the conversions. This disadvantage is also reflected in the fact that the FM function requires a wider range of frequencies near the carrier frequency to achieve the modulation over that required with AM systems. In AM, if the modulating audio signal contains sine waves up to a minimum frequency f_m, then the range of frequencies required for the AM modulation and transmission is $f_c - f_m$ to $f_c + f_m$, where f_c is the carrier frequency. This is illustrated graphically in *Figure 4-4.* The difference between the upper frequency of the required range, which is $f_c + f_m$, and the lower frequency of the range, which is $f_c - f_m$, is called the bandwidth requirements of the system. In the case of AM it is equal to $2f_m$. Thus, an AM system broadcasting audio frequencies up to 10 kilohertz (10,000 cycles per second) on a carrier frequency of 1 megahertz (1,000,000 cycles per second) would require frequencies after modulation of from 990,000 hertz to 1,010,000 hertz for a band width of 20,000 hertz. As shown in *Figure 4-4* this is the bandwidth for broadcast station #2. If similar AM stations are to be broadcasting using a carrier frequency on each side of the 1 megahertz carrier of station #2, they could be placed no nearer than $f_{c3} = 1{,}020{,}000$ hertz on the high side and $f_{c1} = 980{,}000$ hertz on the low side, if the information broadcast by one station is not to interfere with that broadcast by the stations adjacent to it. Thus, the bandwidth requirements of a system are important in determining how many separate systems can use a given band of carrier frequencies. Fifty 20-kilohertz bandwidth system

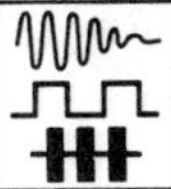

would occupy the frequency band from 1 megahertz to 2 megahertz – a considerable limitation on the utilization of this band of carrier frequencies. However, it is not as severe a limitation as would occur if FM systems were used.

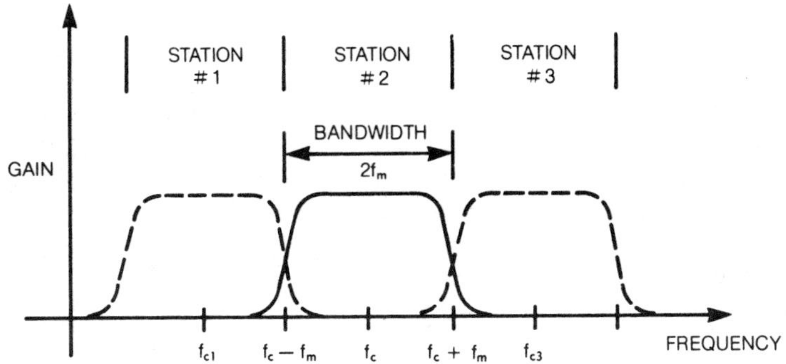

Figure 4-4. *Bandwidth Requirements of an AM Communication System.*

In FM systems the bandwidth is:

$$\text{Bandwidth} = (B + 2)\, 2f_m$$

where B is a characteristic constant of the modulation and is typically between 0.5 and 10. For systems using large B (primarily to reduce noise effects) the FM bandwidth requirements can be many times that of the AM system. For example for a B = 8 system, the bandwidth would be 20 times f_m and with $f_m = 10$ kilohertz as before, the bandwidth is 200 kilohertz. Only five stations could simultaneously use the frequency band from 1 to 2 megahertz without affecting each other, a reduction by a factor of 10 from the number of AM stations. This is why FM station frequencies are in a much higher frequency band, 88 to 108 megahertz. Otherwise, only a few systems would be allowed to operate at a given time in a given area.

Improved Power and Noise Performance

Offsetting the bandwidth disadvantages of FM is its improved power and noise performance over AM systems. Since the carrier contains all the information in its frequency patterns and the frequency of a sinusoid does not affect its power level, all of the FM system power is used in transmitting information. In an AM system only 1/6 of the power transmitted is relative to the information. Thus, the FM system has a 6 to 1 power advantage over the AM system.

One reason FM is chosen over AM when true reproduction of the input information is required at the receiver, as in high fidelity systems, is because noise has much less effect on the FM signal. As a comparison, in the AM system, the signal is transmitted at twice the power relative to the noise; in an FM system the factor is $3B^3$, where B is the same modulation constant as before. When $B=8$, the FM system noise performance is 750 times better than typical AM system noise performance. The reason for this is as follows: the noise that adds to the FM signal during transmission adds to all frequency components equally, affecting mostly their amplitude. Since the FM demodulator only looks at frequency patterns, these amplitude variations are not recognized in the FM system. As a result, an FM system is relatively immune to this type of noise interference. The noise performance of systems can be improved even further by using digital techniques instead of analog signals.

DIGITAL COMMUNICATIONS SYSTEMS

Digital communications, as stated previously, refers to the fact that the information is carried in patterns of binary (two-level) signals, whether the modulation used is AM or FM, or whether the carrier is modulated with straight pulse techniques.

Noise

The advantage that digital signals have over analog signals in terms of their immunity to noise can be seen by examining *Figure 4-5*. Take the case of a digital signal with noise is shown in *Figure 4-5a*. If a threshold is used that is halfway between the 0 and the 1 level, and any voltage above this threshold is recorded as a 1 and anything below this value is a zero, the digital information can be reliably reconstructed (*Figure 4-5b*) as long as the noise amplitude is less than ½ the value of the 1 level voltage minus the 0 level voltage. This same amount of noise with a comparable amplitude analog signal *(Figure 4-5c)* sufficiently masks the small variations in the analog signal so that it is not possible to faithfully reconstruct the analog signal buried in the noise.

Of course, if the digital signal is further encoded using FM modulation, it is even less susceptible to the noise. Also, by using a technique called differential encoding which transmits the signal and its inverted waveform along similar paths, a large part of the noise can be cancelled out, and the signal can be recovered even in the presence of large amplitude noise voltage. It can be applied effectively to both analog and digital signals, though the digital signals would still retain their superior noise immunity over analog signals.

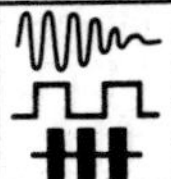

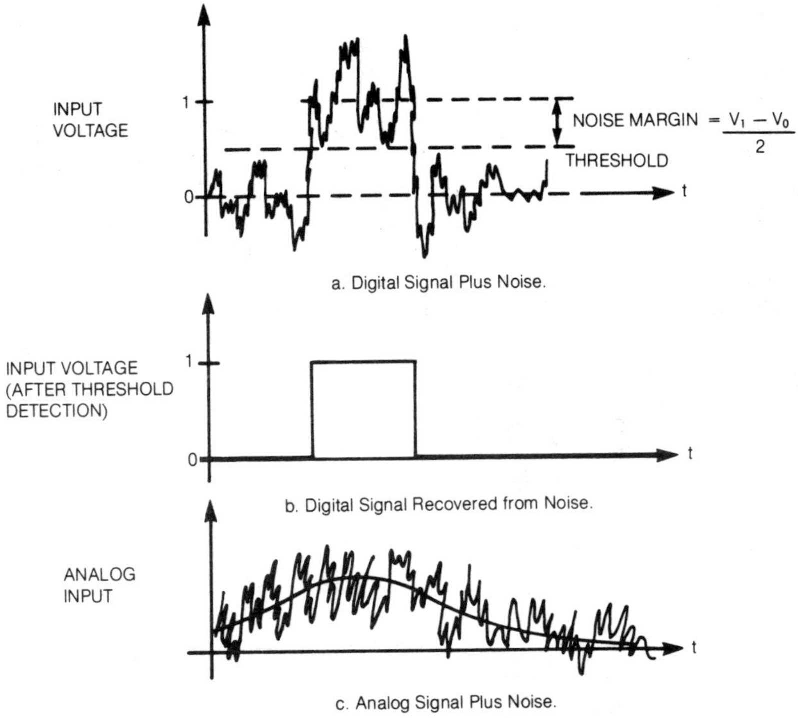

INPUT VOLTAGE

1

0

NOISE MARGIN $= \frac{V_1 - V_0}{2}$

THRESHOLD

t

a. Digital Signal Plus Noise.

INPUT VOLTAGE (AFTER THRESHOLD DETECTION)

1

0

t

b. Digital Signal Recovered from Noise.

ANALOG INPUT

t

c. Analog Signal Plus Noise.

Figure 4-5. *Effects of Noise Voltages on Electronic Signals.*

Main Features

Analog information, when transmitted by using digital system techniques, must be converted to digital signals. Therefore, as shown in *Figure 4-6*, A/D and D/A converters are normally added in digital systems in addition to the modulation, demodulation, and amplification. The modulators and demodulators and the systems from there on could be AM or FM. Any of the previous techniques of PWM, PCM equivalent binary codes can be used for the A/D and D/A functions to obtain a pattern of digital information. Included in the A/D and D/A units of the system of *Figure 4-6* must be a shift register to feed serial data to the modulator and accept serial data from the demodulator. After the D/A or before the A/D conversions, amplifiers are used to increase the signal level of the analog signals. Similar to the AM and FM systems, audio amplifiers are shown in *Figure 4-6* for this purpose.

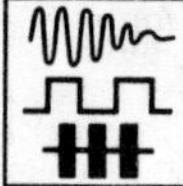

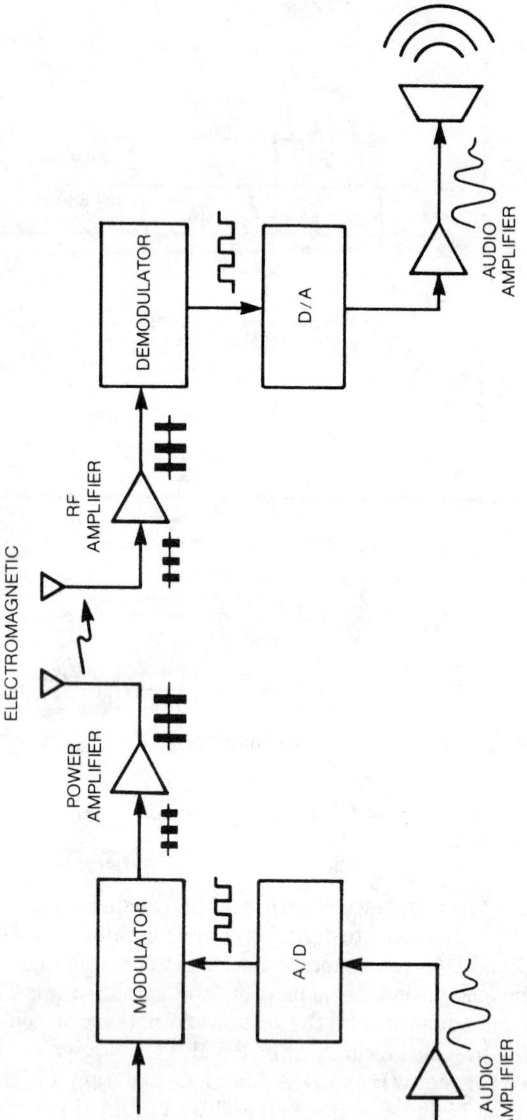

CARRIER
MODULATOR
A/D
AUDIO AMPLIFIER
POWER AMPLIFIER
ELECTROMAGNETIC
RF AMPLIFIER
DEMODULATOR
D/A
AUDIO AMPLIFIER

Figure 4-6. *Basic Structure of Digital Communications System.*

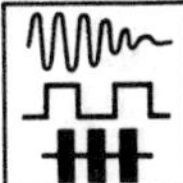

In some cases, when one computer is conversing with another computer over a communications network, the A/D and D/A conversion is not necessary at all, since the information is naturally in digital form. Even so, the system may require conversion from parallel binary data form to serial data form or vice versa. These functions were presented in *Figures 3-16* and *3-17*.

Full featured communications systems are expensive. As more and more electronic communications occur, more and more system hardware is required. Therefore, more efficient ways of using the system are being investigated. One such technique is time-multiplexing wherein the same system is shared to handle a multiple number of communications tasks.

TIME-MULTIPLEXED SYSTEMS

One way to use a single communication system to serve several different communications tasks is to have each communicator take turns, in time. This is really the way the telephone in a home works when there is only a single line into the house and several people want to talk. They have to do it one at a time. As soon as one conversation is over and the phone is hung up, the next conversation can be handled. This continues until all conversations have been handled over the single telephone line.

Any single channel system can be shared by several communicators, each with his own message, by alloting certain times to each communicator. The overall system is illustrated in *Figure 4-7a.* A switch in the transmitter is synchronized to a similar switch in the receiver. The switches start in position 1 and stay there for a period of t seconds, during which time Source 1 is transmitting and Destination 1 is receiving the information. The switches go to position 2 for another period of t seconds, during which time Source 2 and Destination 2 are communicating. The switches go to positions 3 for t seconds, connecting Source 3 to Destination 3. The switches repeat the sequence starting at position 1 allowing the three sets of communicators to time-share the same bandwidth, the overall bandwidth of the system is the same regardless of how many pairs share the channel. The standard type of communication systems can be used from the source switch inputs to the destination switch outputs. But each communicator must wait his turn, which in the case of *Figure 4-7b* would mean a wait of 3t seconds between the start of transmissions. Usually this delay is not a problem except in systems requiring very high speed transfer of information.

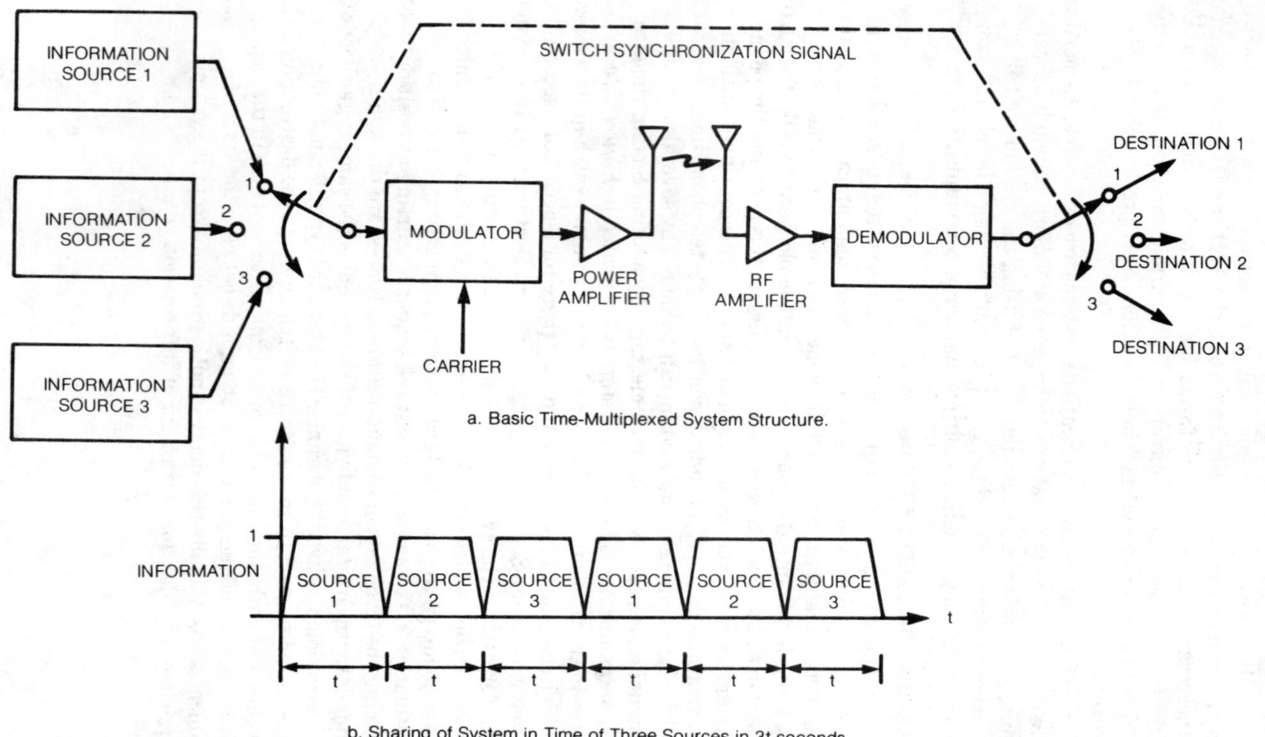

a. Basic Time-Multiplexed System Structure.

b. Sharing of System in Time of Three Sources in 3t seconds.

Figure 4-7. *Time-Multiplexed Communications System.*

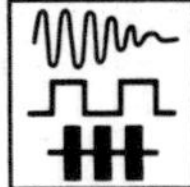

FREQUENCY-MULTIPLEXED SYSTEMS

An alternative way of sharing the same communication system without forcing any of the participants to wait on any others to finish their communication is shown in *Figure 4-8.* In this approach each set of communicators are assigned a range of frequencies, as defined by the bandwidth requirements of their communications. This separates the communicators throughout the frequency spectrum as shown in *Figure 4-8b.* Since none of the bandpasses overlap, none of the communicators interfere with each other. At the transmitter, the information from each source is modulated in such a way that it occupies its assigned range of frequencies. All of these non-overlapping sets of communications signals are summed and sent to the transmitter antenna, using common final modulators and power amplifiers. At the receiver, RF amplifiers increase the signal level of the summed signal, and a common demodulator delivers the signal to the bandpass amplifiers. The bandpass amplifiers separate out each range of frequencies sending the information for Destination 1 out through bandpass amplifier 1, for Destination 2 out through bandpass amplifier 2, and for Destination 3 out through bandpass amplifier 3. All communications can go on at the same time. However, since the overall bandwidth is the number of pairs of communicators times the bandwidth required for each communication pair, the overall bandwidth required of the system is increased over a single user system. In the example of *Figure 4-8,* the overall bandwidth required would be three times that required if only one set of communicators were using the equipment. This would also be true if the communicators were each using separate transmitters and receivers, so the frequency mulitiplexed technique is a much more efficient use of hardware.

Multiplexing of information onto common transmitter-receiver channels is one way of improving the system performance in terms of system costs. There are many techniques of improving system performance in terms of other system parameters which will be discussed in the later chapters. All of these techniques will build on and expand the basic concepts presented in this chapter.

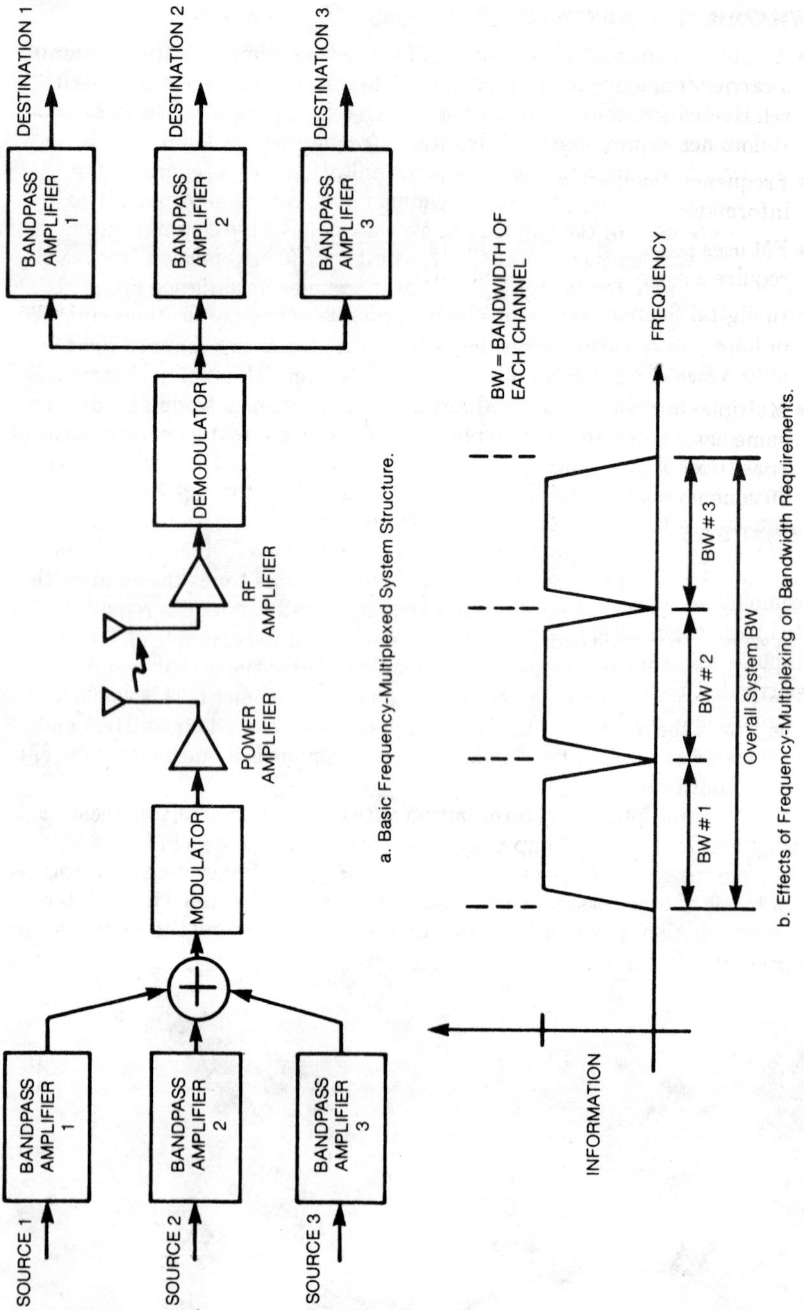

a. Basic Frequency-Multiplexed System Structure.

b. Effects of Frequency-Multiplexing on Bandwidth Requirements.

Figure 4-8. *Frequency-Multiplexed Communication System.*

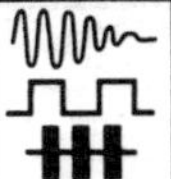

WHAT HAVE WE LEARNED?

- Amplitude modulation involves using patterns of amplitude variations of a carrier frequency to carry information. It is a simple system, though it is relatively inefficient in its use of power and is susceptible to noise effects. It does not require much bandwidth.
- Frequency modulation involves using patterns of frequency to carry information.
- FM uses power efficiently and is fairly immune to noise, though it does require a relatively large bandwidth.
- In digital communications systems, the information is carried in patterns of binary bits. These may be pulse codes in themselves or be modulated onto AM or FM carriers.
- Multiplexing is a method of allowing several communicators to share the same basic transmission and receiver equipment, either by taking turns in time (time-multiplexing) or by using non-overlapping frequency ranges (frequency-multiplexing).

WHAT'S NEXT

The ways and means of basic communications systems are now complete at least from input to transmitter output and from receiver input to information output. What happens from transmitter output to receiver input – the transmission link – is the subject of the next chapter. In particular, the electromagnetic radiation spectrum and the properties and regulation of such radiation are the priority subjects.

Quiz for Chapter 4

1. The purpose of the oscillator in the transmitter of an AM system is to:
   a. provide the modulating signal.
   b. provide enough power to meet transmission system requirements.
   c. provide the carrier that is used in the transmission of the information.
   d. provide a stabilizing signal for the AGC.
2. In an AM system the modulator is the device that:
   a. mutliplies the modulating signal times the carrier signal.
   b. provides the power required by the transmission system.
   c. causes the carrier signal envelope to vary in the pattern of the modulating signal.
   d. All of the above.
   e. a and b above.
   f. a and c above.
3. The demodulator in an AM system:
   a. is a low pass amplifier.
   b. is an envelope detector, i.e. it extracts the carrier envelope variations from the incoming modulated signal.
   c. derives the modulating signal from the received signal.
   d. All of the above.
   e. None of the above.
4. If a modulating sinusoid of 10 kHz is used in a double sideband modulation of a 100 kHz carrier sinusoid, what frequency sinusoids are present in the output of the modulator:
   a. 10 kHz
   b. 100 kHz
   c. 90 kHz
   d. 110 kHz
   e. 200 kHz
   f. All of the above.
   g. b, c, and d above.
5. If the modulating signal has a bandwidth of 20 kHz in an AM system, what is the overall bandwidth required of the transmitter?
   a. 20 kHz **c.** 80 kHz
   b. 40 kHz **d.** 10 kHz
6. FM has the following advantages over AM:
   a. better utilization of transmitter power.
   b. better noise rejection.
   c. simpler modulating and demodulating circuits.
   d. a and b above.
   e. a and c above.
7. In an FM system with modulating frequencies up to 20 kHz and a total bandwidth of 200 kHz, the signal is boosted over noise by:
   a. 3 times. **c.** 36 times.
   b. 10 times. **d.** 81 times.
8. If a digital 1 level is 5 volts and the digital 0 level is 1 volt, what is the noise margin in volts?
   a. 1 **c.** 3
   b. 2 **d.** 4
9. Time-multiplexing allows:
   a. several subsystems to share the same transmitter facilities.
   b. different time periods to be used by different communicators.
   c. only one modulating frequency to be used to handle many communicators.
   d. All of the above.
10. Frequency-multiplexing allows:
    a. several subsystems to share the same transmitter facilities.
    b. different frequency ranges to be used by different communicators.
    c. several communicators to carry on communications simultaneously using the same transmitter.
    d. All of the above.

(Answers in back of the book)

The Communications Spectrum

ABOUT THIS CHAPTER

Previous discussion has shown how the properties of electrical signals can be varied in order to carry information from one place to another. Understanding the ways and means in which information is represented as patterns of these properties, transmitted, received and converted back to the original information resulted in a general overview of the paths used to send information from the transmitter to the receiver. Several alternatives were shown. The transmission link could be such that the electrical energy is carried through space in the form of electromagnetic radiation or through wires or through paths that include amplifiers. The choice of the path the electrical signal can take in carrying the information depends on the frequency of the signal. The entire range of frequency alternatives available is called the communications spectrum. The relationships between the spectrum and the manner in which electrical energy is transferred from one place to another will be the subject of this chapter.

WHAT IS MEANT BY SPECTRUM?

Spectrum is the entire available range of sinusoidal signal frequencies. The signal pattern of energy could be sound, electrical, or some other physical form. In this book the emphasis is on electrical signals since the thrust of the book is electronic communications. Fortunately, many of the concepts encountered in this subject have familiar examples that everyone can relate to. Spectrum is one of these concepts, since everyone has visually experienced the concept of spectrum through the familiar rainbow.

When one views a rainbow, they are viewing the entire range of visible electromagnetic radiation frequencies at one time. They are viewing a spectrum of frequencies of light energy. To the human eyes these are interpreted as colors ranging from the deep purple to the deep red. If the human eye were sensitive to all frequencies of electrical signals in the form of electromagnetic radiation, the rainbow would simply be broader, and the "color" range a much wider variety. Of course, it is not possible to conceive of colors other than those that exist in terms of human visual capabilities, but the "feeling" of the spectrum that the rainbow gives with its spread of colors is a very direct understanding of the concept of spectrum. Now the color band experienced by the human eye is a very small portion of the upper end of the entire electromagnetic spectrum. The frequencies of visible electromagnetic signals that can be viewed and their corresponding wavelength is shown in *Figure 5-1*.

$$f\lambda = 300{,}000{,}000 \text{ meters/second (speed of light)}$$

$$\lambda = \frac{300{,}000{,}000}{f}$$

$$\text{wavelength (in meters)} = 300{,}000{,}000/\text{frequency (in cycles per second-hertz)}$$

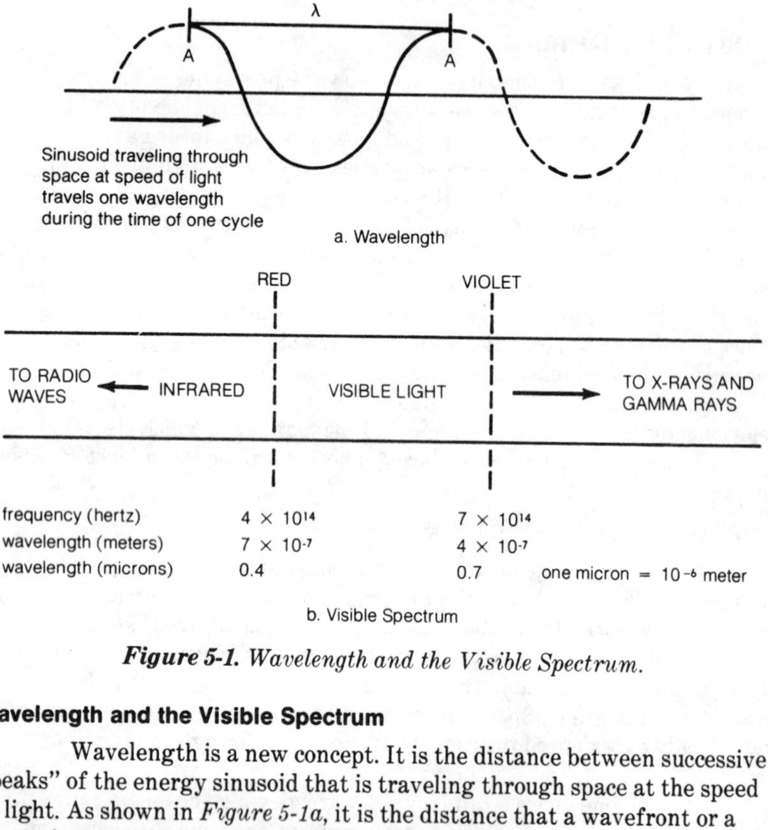

a. Wavelength

	RED	VIOLET	
frequency (hertz)	4×10^{14}	7×10^{14}	
wavelength (meters)	7×10^{-7}	4×10^{-7}	
wavelength (microns)	0.4	0.7	one micron = 10^{-6} meter

b. Visible Spectrum

Figure 5-1. *Wavelength and the Visible Spectrum.*

Wavelength and the Visible Spectrum

Wavelength is a new concept. It is the distance between successive "peaks" of the energy sinusoid that is traveling through space at the speed of light. As shown in *Figure 5-1a*, it is the distance that a wavefront or a given sinewave frequency, traveling at the speed of light, travels in space in the time of one cycle. The speed of light is 300,000,000 meters per second. The relationship between the frequency in hertz and the wavelength in meters is simply that the frequency times the wavelength is equal to the speed of light. In *Figure 5-1b* this wavelength is listed in meters as well as in the more convenient unit of microns (with one micron being one millionth of a meter or 10^{-6} meters). *Figure 5-1b* illustrates that the visible spectrum of frequencies is from 4×10^{14} hertz to 7×10^{14} hertz. The frequencies lower than the visible radiation range are where the radio waves used in conventional electronic communications occur. Modern electronic communications systems use the entire range of frequencies all the way from a few hertz to, and including, the visible light range.

The Sound Spectrum

Another type of spectrum that humans can experience directly occurs in the band of frequencies which carry sound energy. Human ears are sensitive to the range of frequencies from a few tens of hertz up to above 10,000 hertz. Sound energies above 10,000 hertz may not be heard by most humans but many animals can detect these sounds. Therefore, the full spectrum from less than 1 cycle per second to 20,000 hertz can be used to carry information. Everyone who can hear has a feeling for how the features of this sound spectrum vary from the very low bass notes of musical instruments, through the human voice range, on up to the high pitched notes and overtones of the trumpets, violins, and piccolo musical instruments. *Figure 5-2* shows the relationship between frequency, the piano keyboard, and familiar musical notes for the sound spectrum. Sound travels much slower through the air surrounding us than electromagnetic waves. Here the wavelength times the frequency is equal to the speed of the form of energy, which is the speed of sound, about 741 miles per hour at sea level. There are also electrical currents and electromagnetic radiation signals at these same frequencies that have been used in electronic communications systems, but their wavelengths are around a million times longer than the corresponding sound energy wavelength would be.

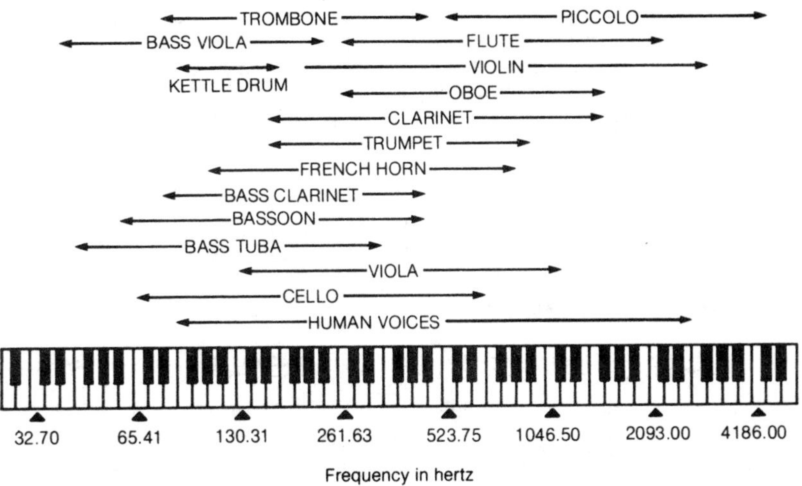

Figure 5-2. *The Sound Spectrum.*
(Source: Reference Data for Engineers-4th Ed. 1968, 1975, H. W. Sams Inc.)

Both of these familiar spectrums of visible radiation and audible sound waves allow man to comprehend in a very intuitive way some concepts that otherwise would be very abstract. Throughout this chapter the human experience and feel for these forms of physical spectra will be used to help understand the features and behavior of the electrical energy used in electronic communications systems. This will be particularly helpful in understanding the nature of electromagnetic radiation and how its properties vary with frequency and energy. A helpful chart to convert from frequency to wavelength for the electromagnetic wave spectrum, both in meters and in feet, is shown in *Figure 5-3.* Applying the multiplication factors allows coverage of any portion of the spectrum.

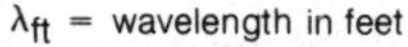

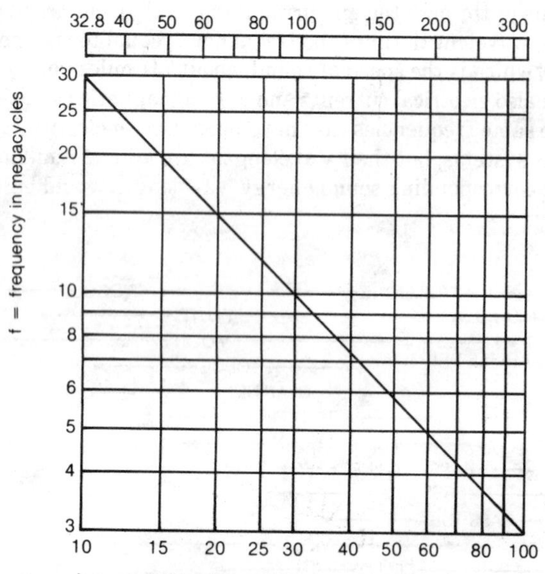

for frequencies from	multiply f by	multiply λm by
0.03 — 0.3 megacycles	0.01	100
0.3 — 3.0 megacycles	0.1	10
3.0 — 30 megacycles	1.0	1.0
30 — 300 megacycles	10	0.1
300 — 3,000 megacycles	100	0.01
3,000 — 30,000 megacycles	1,000	0.001
30,000 — 300,000 megacycles	10,000	0.0001

***Figure 5-3.** Conversion of frequency to Wavelength for Electromagnetic Waves.*

(Source: Reference Data for Engineers-6th Ed. 1968, 1975, H. W. Sams Inc.)

What are the Properties of Electromagnetic Radiation?

Electromagnetic radiation can occur when electrical currents of sufficient amplitude flow in an antenna whose dimensions are about the size of the wavelength of the radiation to be generated. In fact, radiation can be generated by devices that are not supposed to produce radiation, such as receivers or industrial machines. This unwanted radiation can cause interference with radiation that is actually carrying information. When the radiation is produced by a radio transmitter, the pattern of energy depends primarily on the design of the antenna. All of the possible radiation patterns have counterparts in visible patterns that can be seen radiating from sources that are familiar to everyone.

Radiation Patterns

The most obvious pattern of distribution of radiated energy is a uniform one in which the amount of energy is the same in all directions from the transmitter. The more powerful the transmitter, the farther these energy patterns extend in space. This situation is similar to the water sprinkler shown in *Figure 5-4* that puts out a spherical pattern of water, uniformly watering the lawn in all directions from the sprinkler – the higher the water pressure, the larger the area watered. A more direct analogy is that of the electric light bulb shown in *Figure 5-5.* It emits equal amounts of electromagnetic radiation in the form of visible light in all directions from the bulb. The brighter the light (the more electrical energy used to produce the light) the farther away it can be seen.

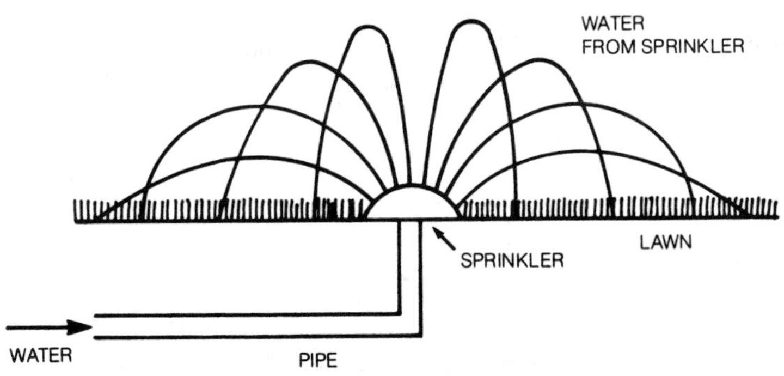

Figure 5-4. *Spherical Radiation Pattern of Water Sprinkler.*

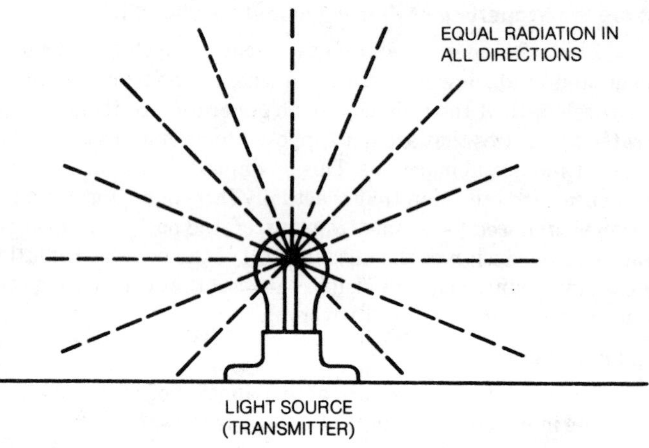

Figure 5-5. *Visible Light Radiation Pattern.*

This type of uniform radiation pattern is needed if the receiver location could be anywhere in space. However, if the receiver location is fixed, this type of pattern is very undesirable, as an examination of *Figure 5-6* will show. The receiver in this case is a photodetector and amplifier. The receiver lens intercepts only a small portion of the total light pattern coming from the transmitter. This is the only energy useful in transmitting information. All of the rest of the energy is wasted, and in fact, may be detrimental because unwanted energy is being sent to other receivers that are not communicating with the transmitter shown in *Figure 5-6*. To avoid such possible interference with other communication systems and to avoid wasting energy, in many cases a more directional pattern is needed.

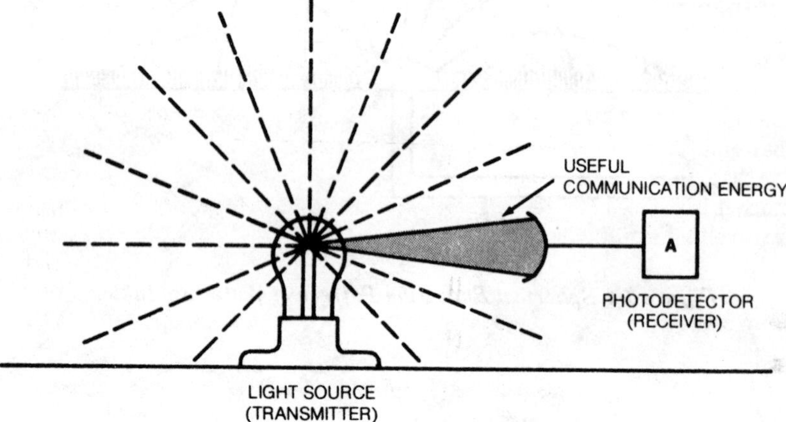

Figure 5-6. *Fixed Location Receiver in Uniform Pattern.*

Directional Radiation Patterns

If the antenna can be designed so that the radiation is emitted as a beam of energy, the energy can be directed to the receiver, and all energy is used to send the information to be communicated. The comparable device in water sprinklers is the spray nozzle which allows the water to be directed in a narrow, forceful stream at one area of the lawn. The familiar flashlight of *Figure 5-7* is an example of a light source that has a beam type of radiation pattern. All of the energy produced by the transmitter is directed in a more concentrated pattern and much more of the radiated energy is able to be collected by the receiver.

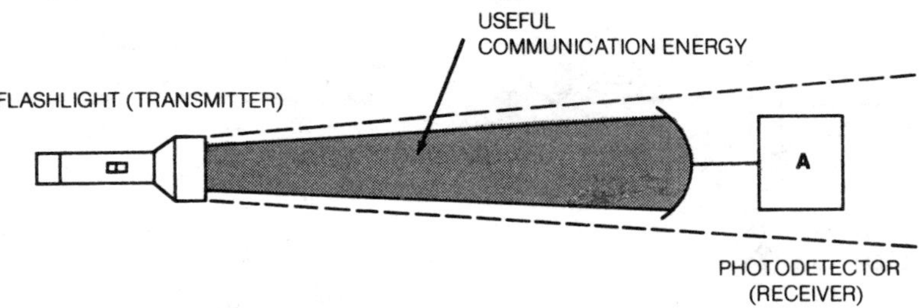

Figure 5-7. *Beam Pattern of Electromagnetic Radiation.*

With such beam radiated patterns, several similar transmitter-receiver pairs can operate in the same area without interfering with one another. This is shown in *Figure 5-8*. It's almost as if the radiation were being conducted in "air pipes" so that each communication system is entirely isolated from all others. There is no chance of the information from system 1 being received by either of the other two systems and vice versa. Again, the stronger the transmitter beam, the further away the receiver can be placed and still provide reliable communications. This assumes that the electromagnetic radiation is not affected by the characteristics of the material through which it is travelling (propagating). Generally, this is not the case. How the electromagnetic radiation travels through space and how much it is reduced in strength, depends on the atmosphere through which it is travelling and the frequency of the radiation.

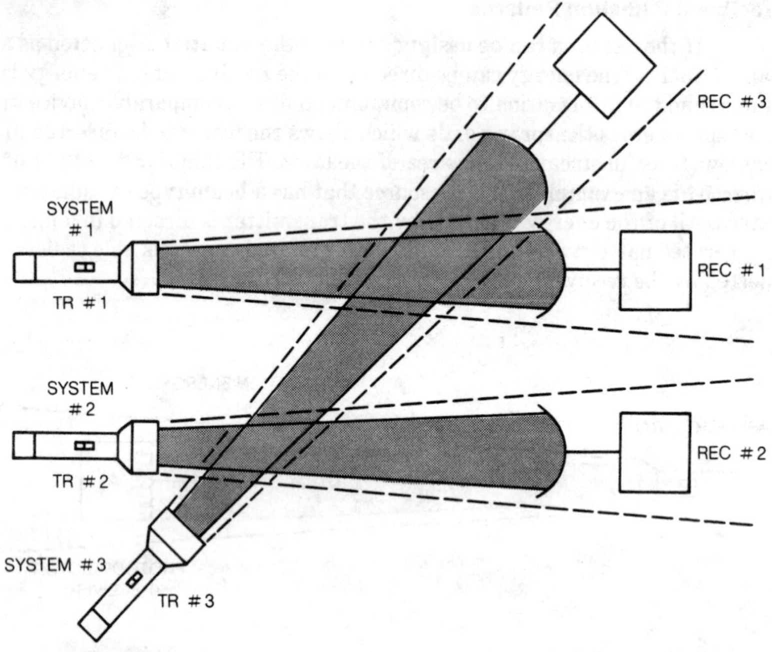

Figure 5-8. *Electromagnetic Beam Communication Systems.*

Electromagnetic Wave Propagation

Two things affect the path and distance travelled by the radiated electromagnetic waves. One is the weather, because it affects the atmosphere through which the wave travels. Another is the carrier frequency of the radiated waves. Using the flashlight analogy again, it is difficult to get a well directed flashlight beam from one point to another on a very foggy night, even with a very powerful light beam, because the light energy is being reflected by the particles in the air. Radio waves have like problems with rain, snow, sleet and similar weather effects. They may be reflected, or lose energy, or be dissipated depending on atmospheric conditions and on the frequency.

<u>ELF and VF Bands</u>

In order to summarize these effects, it will be useful to introduce the common terminology used to describe the various bands of radio frequencies. This terminology is summarized in *Figure 5-9*. The extremely low frequency ELF band contains many frequencies produced by musical instruments and the human voice as shown in *Figure 5-2*, however, the voice frequency band VF, contains a major part of the energy spectrum of human speech.

VLF and LF Bands

The very-low frequency (VLF) and low frequency (LF) bands were initially used for radio telegraphy. Since the wavelengths are in the kilometer range and higher (30kH would have a $\lambda = 10$ Kilometers, or about 6.2 miles), enormous antennas have to be used to generate radiation in this range. As a result, such frequencies are used for special applications in modern communications systems.

band number	frequency range		metric subdivision	frequency subdivision	
2	30–	300 H	Megametic waves	ELF	Extremely low frequency
3	300–	3,000 H	———	VF	Voice frequency
4	3–	30 kH	Myriametric waves	VLF	Very-low frequency
5	30–	300 kH	Kilometric waves	LF	Low frequency
6	300–	3,000 kH	Hectometric waves	MF	Medium frequency
7	3,000–	30,000 kH	Decametric waves	HF	High frequency
8	30–	300 MH	Metric waves	VHF	Very-high frequency
9	300–	3,000 MH	Decimetric waves	UHF	Ultra-high frequency
10	3,000–	30,000 MH	Centimetric waves	SHF	Super-high frequency
11	30,000–	300,000 MH	Millimetric waves	EHF	Extremely-high frequency
12	3,000,000–	3,000,000 MH	Decimillimetric waves	—	—

Figure 5-9. *Terminology of Frequency Bands.*

(Source: Reference Data for Engineers-6th Ed. 1968, 1975, H. W. Sams Inc.)

MF and HF Bands

The medium frequency (MF) and high frequency (HF) bands are commonly used for commercial AM broadcasting as well as for short wave and amateur radio operations. Electromagnetic radiation in these frequency ranges exhibits the important property of being reflected by the ionosphere. This reflection and its effect are illustrated in *Figure 5-10.* The ionosphere is a layer of electrically charged particles at the top of earth's atmosphere. The layer is caused by the strong solar radiation entering the upper atmosphere. When an electromagnetic wave in the MF or HF range hits this layer it is reflected back to earth. Multiple reflections between this layer and earth are possible, allowing great distances (even around the world) to be obtained by transmitters operating in these frequency ranges. This is particularly true for the high frequency band.

The disadvantage of this type of propagation is that it depends on the characteristics of the ionosphere, which vary widely, especially during the day hours. As a result, the waves are reflected differently and take different paths over a period of time, causing the signal at a receiver at a particular location to vary in strength. This causes the receiver output to fade in and out.

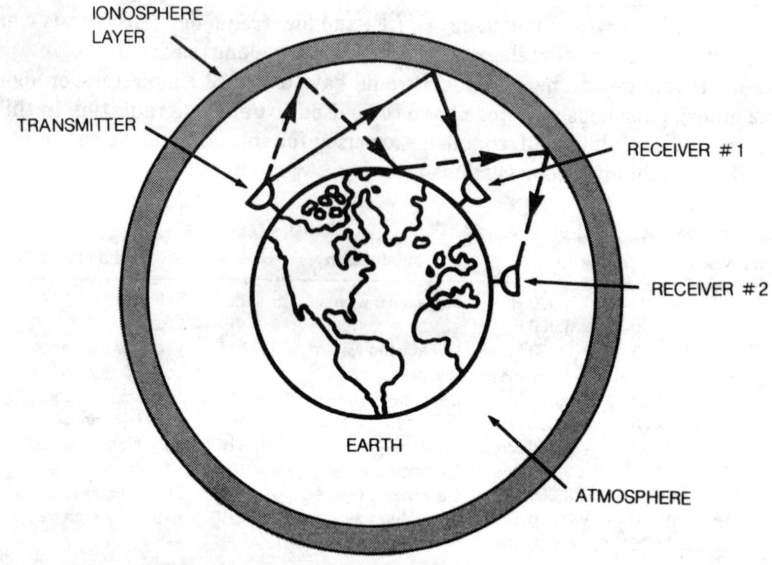

Figure 5-10. *Ionospheric Reflection of MF and HF radiation.*

With medium frequency waves (used by American AM broadcasting) there is little daytime reflection from the ionosphere so the distance of the radiation is limited to 100 kilometers or so, and there is little fading. At nighttime, the ionospheric reflections occur and reception can be obtained at thousands of kilometers with a steady signal, though fading can occur in the 100 to 200 kilometer range. The HF signals, on the other hand, depend a great deal on ionospheric scattering all the time, with the most stable signals occurring at night. As a result, with reasonable power, worldwide communications can be maintained in the high frequency band most of the time.

VHF and UHF Bands

For electromagnetic radiation with frequencies in the very-high frequency band and above, the ionospheric reflection is very small, and therefore, usually not important. As a result, communications with transmitters in this frequency range tend to be line-of-sight and over short distances. Line-of-sight communication is what the name implies, the receiver and transmitter must be within a straight visual sighting line of each other. Buildings and uneven terrain affect the transmission, particularly for the ultra-high frequency UHF band and above. The lower part of the UHF band and the VHF (very-high frequency) band is useful for mobile communications and television.

Microwaves

Everything above the frequencies of one billion hertz (called 1 gigahertz, or 1 GHz) are called microwave frequencies. They are commonly used for radar and wideband communications. One feature of these higher frequency bands is that their radiation can be directed into very narrow beams of energy, much like the flashlight examples of *Figures 5-7.* This makes these ranges very efficient in utilization of transmitter energy as well as minimizing interference between one communication system and another. When the microwave frequency band is from 3 GHz to 30 GHz then the wavelength is from 10 centimeters to 1 centimeter, above 30 GHz it is millimeters in length. This is a significant advantage for transmitting in these frequency bands because the antennas can be very small. As evidence, *Figure 5-11* shows a typical transmitter-receiver antenna relay tower. However, because wavelengths are small, it can be a disadvantage. Transmissions at these frequencies are very susceptible to weather effects, particularly rain, because each rain drop can become a small antenna in itself, absorbing the energy and causing it to be dissipated rather than reaching its destination.

Figure 5-11. *A Repeater Station.*
(Courtesy of Southwestern Bell Telephone Co.)

Effects of Weather on Radiation

When transmissions are at frequencies where they are susceptible to weather, the water and oxygen molecules in the air absorb the electromagnetic energy, making the air more and more opaque to radiation in the millimeter and shorter wavelengths. Under heavy rain the loss of energy is particularly severe at these frequencies. The communications system designer has to take these losses into account when working in these frequency bands. As a result, the distance between transmitter and receiver must be limited to a few kilometers when using microwave frequencies in the SHF band above 10 GHz. Below 10 GHz the spacing between transmitter and receiver can be a few tens of kilometers as determined by the line-of-sight considerations. If the distance of a microwave system is greater than these minimum distances of a few kilometers, repeater or relay stations such as shown in *Figure 5-11* must be placed along the path as shown in *Figure 5-12*.

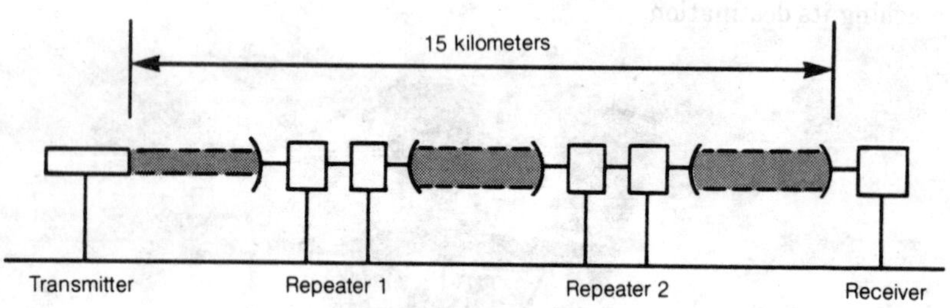

Figure 5-12. *Typical Repeaters Required of a 20 Ghz System.*

The repeater is simply a receiver and transmitter that receives the signal from the previous repeater, boosts its power, and transmits it to the next repeater or the final destination of the information, as the case may be. As the frequencies increase, the distance between repeaters decreases. Proportionately, the size of antennas would also decrease. However, the repeaters are not just for the millimeter and centimeter wavelengths. Even the longer wavelength microwave transmissions may need repeaters if the overall distance between transmitter and receiver exceeds around 40 kilometers. These repeaters are much larger than those used in millimeter wave systems because of the much larger antennas and greater power requirements.

One way to eliminate the disturbing atmospheric affects due to oxygen and water molecule absorption of the radiated energy is to avoid sending the signals through the atmosphere. An alternative is available that offers several advantages to the communications system designer. Closed containers or cables can be used to carry the transmitted information. These closed transmission guides and cables can take any of the several forms depending on the frequency band being used for communications.

HOW IS ELECTRICAL ENERGY TRANSMITTED THROUGH CABLES?

Waveguides

Total control over the transmission of the electrical energy from the transmitter to the receiver is possible if the properties of the path can be controlled at all points. This can be done in the case of microwave electromagnetic radiation propagation by constraining the wave to a metallic pipe, much as water is constrained to a water pipe. When rectangular or cylindrical pipes are used to contain an electromagnetic wave for the purpose of transmission of the wave from one point to another, they are called waveguides. The operation of such a waveguide is illustrated in *Figure 5-13*. The wave cannot escape the pipe because the metallic or reflecting walls contain the energy. Under most conditions the wall reflections are not important and the wave simply travels down the pipe at the speed of light. By controlling the atmosphere in the pipe so that it contains very few oxygen atoms the energy travels down the pipe with little loss. This is a significant advantage. It saves repeater stations because now repeaters can be spaced every 30 or more kilometers instead of 5 to 10.

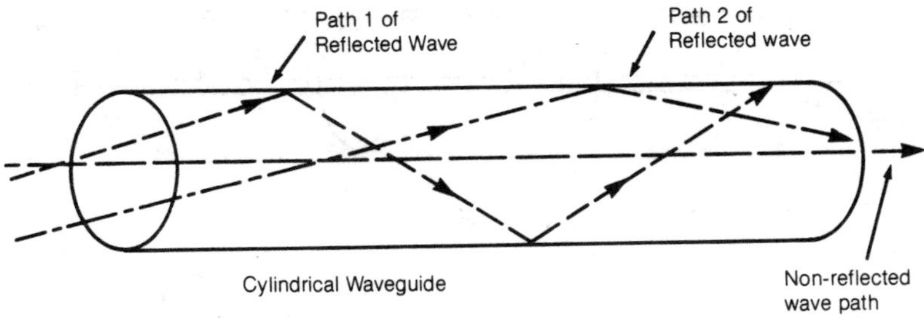

Figure 5-13. *Electromagnetic Wave Waveguide Structure.*

Ten times fewer repeaters would be needed for such a waveguide system at 40 GHz than would be needed for an open air system. Another advantage of the waveguide system is that no outside or unwanted electromagnetic energy can interfere with the information being transmitted. All external radiation is blocked by the metal walls of the waveguide, so that very reliable and dependable communications can be insured by using such a system. The primary frequency bands for use of waveguides are the SHF and EHF bands. Much of this has to do again with the size of the wavelengths and thus the size of the waveguides. Waveguides are quite expensive and their use adds significantly to system cost.

Coax Cables

At frequencies in the VHF range and lower, wires can be used to transmit electrical signals from the transmitter to the receiver. A cable of the structure of *Figure 5-14*, called a coaxial cable, is used for this purpose in most long distance situations. The "coax" cable consists of an inner wire held in position within an insulating medium by circular insulating spacers. Around this is a metal covering. The outer covering prevents signals from other wires or from electromagnetic radiation from affecting the information carried within the coaxial cable. Electrical signals in the frequency range above 100 MHz (100 million hertz) cannot be effectively transmitted through such cables because the dielectric losses and reflected losses are too great.

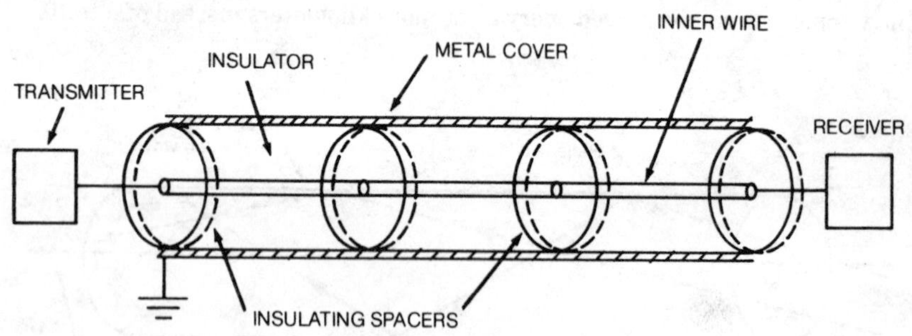

Figure 5-14. *Transmission of Electrical Signals Using Coaxial Cable.*

Fiber Optics

For very-high frequency electromagnetic radiation in the infrared and visible light range, a cable that is similar to the metallic waveguide is used. In this case it is a glass fiber that causes very little loss in the light energy over long distances. The diameter of the fiber must be small in order to minimize the reflective type of transmission shown in *Figure 5-13.* Ideally, all of the light should travel straight down the fiber. This insures the highest speed transmission and makes sure that sharply defined input light pulses stay sharply defined until they reach a receiver.

The basic construction is shown in *Figure 5-15.* A small diameter low-loss glass of relatively high refractive index is used for the central transmitting fiber. This is surrounded by a higher loss, lower refractive index glass for support and for absorbing any rays that leave the central transmission fiber. The light source at the transmitter could be a light emitting diode (an LED) or a laser (a source of higher power, single frequency, narrow beam radiation). The detector at the other end is a photodiode or a phototransistor. These devices can be made relatively inexpensively and can be made to operate at very high speeds for producing high bit-rate digital transmission.

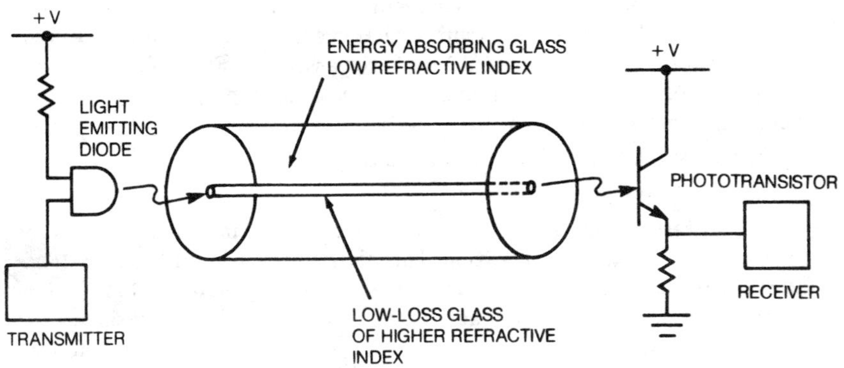

Figure 5-15. *Basic Structure of Optical Fiber Transmission.*

Whether the communication system uses radiation through space or cable connections between the transmitter and receiver depends on the type of communication system being built. Similarly, the frequency bands being used depend on the requirements of the communications problem being solved. Most of the frequency choices available are dictated by national regulations and international agreements. These regulations define how the spectrum is to be used.

HOW IS THE SPECTRUM USED?

FCC

Just how the spectrum is used is determined in the United States by the Federal Communications Commission. The precise rules and regulations can be obtained from the FCC at a nominal charge. These FCC regulations are in turn consistent with all international agreements on the use of the spectrum. While it may seem that the spectrum is an inexhaustable resource, this is not necessarily true. At any given time the technological capabilities of man limits the frequency ranges he can economically use. This limits the resource and fixes the number of users who can use a given part of the spectrum at a given time. This is particularly true for the U.S. AM broadcast band which is located from 535 kHz to 1605 kHz. In any given area the frequencies of stations must be placed far enough apart that they do not interfere with one another. This fixes the number of stations that can be licensed in any area.

The same situation exists in the FM broadcast band which in the U.S. is from 88 MHz to 108 MHz. Everyone who has listened to radio has a fairly good feel for just how close two powerful stations can be in frequency and still not interfere with each other. This is why it is so important to have an agency like the FCC. This independent government agency slices up the pie of the electromagnetic spectrum so that everyone has an equal access to this natural resource. After regulations are passed, the FCC polices the use of the spectrum to make sure that everyone is using it properly without violating someone else's right to it.

Allocation of the Electromagnetic Spectrum

International Regions

International treaties have divided the world into three regions for the purpose of frequency allocation. This division is shown in *Figure 5-16*. Each region has been assigned certain frequency ranges for certain purposes. Some of these assignments are made on a worldwide basis to enable international carriers to use the same equipment everywhere and to provide for communications between the regions. The United States is in region 2, and the FCC makes its frequency assignments so that they are consistent with the international requirements for region 2. All of these regulated allocations apply only to transmitters emitting electromagnetic radiation into space. Cable and waveguide distribution systems that are totally self-contained and do not radiate from antennas may use any frequency for any purpose they choose.

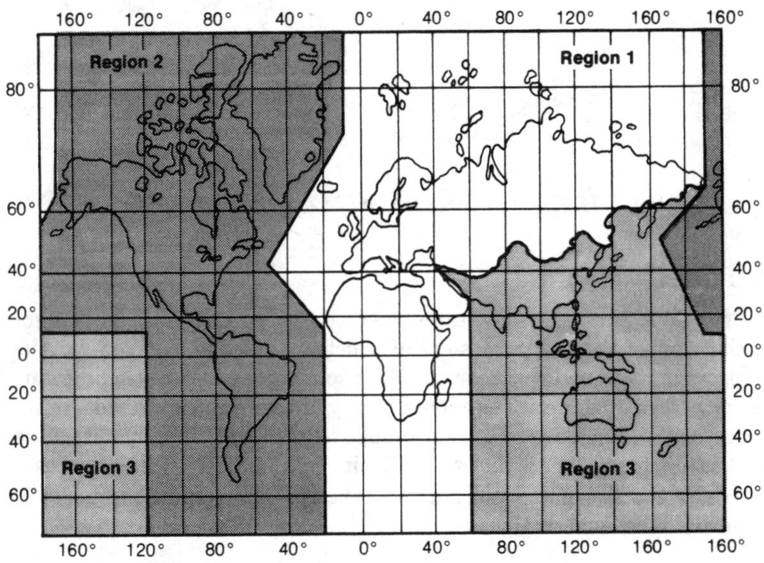

Figure 5-16. *Definition of Regions for Using the International Spectrum.*
(Source: Reference Data for Engineers-6th Ed. 1968, 1975, H. W. Sams Inc.)

International Allocations

The national and international allocations are divided according to categories of users. In the international allocations these categories are rather broad (*Figure 5-17*). One for broadcasting, one for navigational beacons, one for amateur, and one category with four subdivisions for the working communications area. Of particular interest to non-industrial and non-government users is the amateur category, whose frequency allocations are made on a worldwide basis and with ranges of frequencies in every important band of the spectrum. This is very important since it gives every citizen of the world free and fair access to the electromagnetic spectrum for his own use. In the United States the FCC decides under what conditions a citizen may actually use these amateur frequencies (or any other frequency band for that matter).

Broadcasting
Navigational Beacons
Amateur
Fixed or Mobile; Land, Maritime, or Aeronautical

Figure 5-17. *General International Categories of Spectrum Use.*

FCC Categories

Within the United States the FCC defines fairly specific categories of the users of the spectrum, some of which are listed in *Figure 5-18.* Almost all users in these categories must obtain some type of license in order to operate electromagnetic equipment in any of the frequency bands. In some cases these licenses are relatively simple, involving only filling out a license application form. This is the case with citizen's band radio. In other cases, extensive engineering exams must be taken to certify radio engineers for radio and television stations.

Broadcasting

- AM
- FM
- TV
- Links
- Common Carriers

Citizens Radio

- Amateur
- CB

Government and Industrial Communications

- Armed Services
- Governmental Departments
- Public Safety
- Industrial Communication
- Meteorological Services
- Telemetry
- Industrial, Scientific, and Medical Equipment

Commercial Transportation Communications

- Paging Services and Radiotelephones
- Mobile Land Vehicles
- Aeronautical Control
- Aviation
- Maritime
- Navigational Beacons

Figure 5-18. *FCC Spectrum Categories—Example Listing.*

Broadcasting

Most of the categories of *Figure 5-18* are relatively self-explanatory. The services of broadcast stations are authorized and controlled through the broadcast category. Everyone uses it daily. The frequency ranges assigned to these services are shown in *Figure 5-19*. In the United States the AM band occupies the frequencies from 535 kHz to 1605 kHz. The FM band occupies the frequencies from 88 MHz to 108 MHz. The VHF television band occupies two ranges of frequencies from 54 to 72 and 76 to 88 MHz and from 174 to 216 MHz. The range of frequencies assigned to UHF television is 470 to 806 MHz.

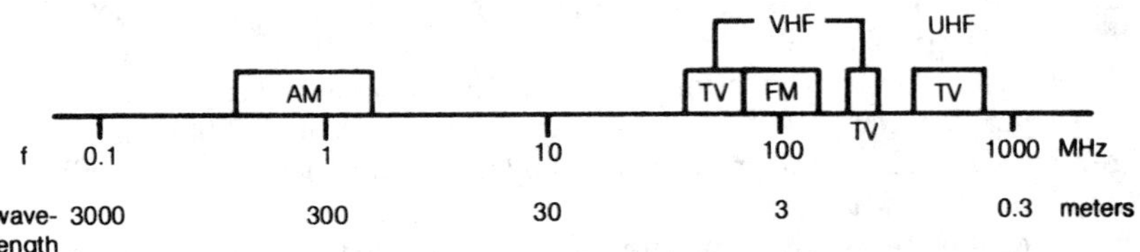

Figure 5-19. *United States Broadcast Bands.*

For those in the broadcast industry, there are also frequencies assigned to enable the studio to control a remote transmitter, (studio-to-transmitter links) and for intercity and continental transmission of programs using microwave relay links and telephone system links. Usually these frequencies are of interest only to radio engineers involved in the details of operating broadcast stations.

Citizen Communication

CB

There are several frequency bands assigned for the use of ordinary citizens for personal communications and experimentation. These are the familiar CB, citizens band, and the amateur bands. These are shown graphically in *Figure 5-20*. The citizens band is dedicated to personal communications and control activities. It uses the frequencies from 26.96 to 27.23 MHz and the 27.255 MHz frequency. This is the band used by the 40-channel commercially available CB radios and transceivers (transmitter and receiver). Garage door openers typically operate in this range of frequencies. There is a band of frequencies of 72 to 76 MHz for use by hobbyists for remote control of airplanes and other models. Finally, there is a special band from 460 to 470 MHz for short distance communication by citizens. All of these frequencies may be used by anyone willing to buy the relatively inexpensive communication equipment.

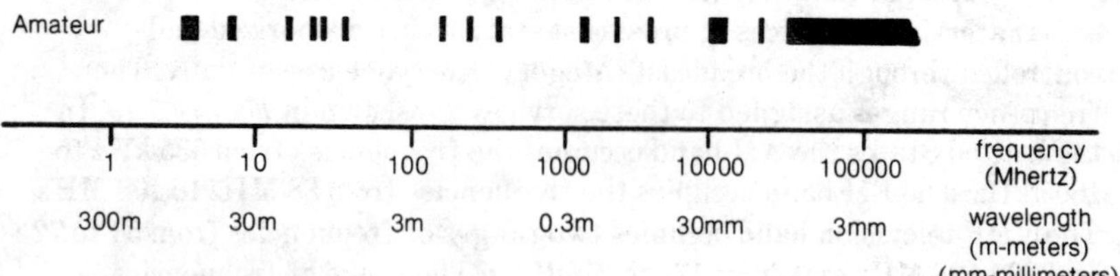

Figure 5-20. *CB and Amateur Bands.*

Amateur

For those non-commercial individuals that want more capability and variety than the CB frequencies offer, there are the amateur radio frequencies graphically shown in *Figure 5-20* and listed by frequency in *Figure 5-21.* These frequencies span the entire range of the electromagnetic spectrum. Amateurs use these for personal conversations within their region, across the nation, and even around the world. They are continually experimenting with communications techniques, continually evaluating equipment of new design, and offering contributions to the art and science of communications. During emergencies they offer their services to set up emergency communications networks. Often the only communications with people in an area struck by a hurricane or other natural disaster is through amateur radio operators. These contributions and the potential for other contributions from citizens around the world who possess their own personal communications equipment have been considered important enough by the governments around the world that these governments have made sure that amateurs have had available frequencies in every important band of the spectrum. It is quite likely that much of the groundwork in the area of adapting satellite communications to home applications will be performed by interested amateurs.

Frequency (MHz)	Common Band Wavelength (meters)
3.5-4.0	80
7-7.3	40
14-14.35	20
21-21.45	15
28-29.7	10
50-54	6
144-148	2
2300-2450	0.12
3300-3500	0.085
5650-5925	0.053
10000-10500	0.03
21000-22000	0.015
Above 30000	0.011

Figure 5-21. *Listing of U.S. Amateur Radio Bands.*

Government and Industrial Communications

A large number of frequencies are assigned to government and industrial communications activities. The specific frequencies are of little interest to those not involved in the design or operation of these systems. The applications include local and worldwide communications for government, armed services, and meteorological services. Local use of the spectrum by factories, businesses, and public safety organizations for communications and control and use in industrial, scientific, and medical equipment are fairly commonplace in almost all areas of the world.

Commercial Transportation Communications

Mobile radio is used for taxicabs, trucks, and other land commercial vehicles. Paging services and radiotelephones also fall into this category. The communications carried on with aircraft, ships, and land transportation vehicles such as buses and trains utilize frequency ranges across the entire spectrum. Frequencies are also allocated to provide coded beams of electromagnetic waves to serve as navigational beacons for aircraft and ships.

Each of the categories of applications of the electromagnetic spectrum have their own requirements of how much of the spectrum they need. These requirements often fix the band that a given user will utilize.

WHAT ARE THE FREQUENCY REQUIREMENTS OF TYPICAL COMMUNICATIONS SYSTEMS?

The frequency required for an application is usually the most important factor that determines how many users can simultaneously use a range of frequencies in the electromagnetic spectrum. The other factor is the transmission pattern used by the system. A system that broadcasts equal power in all directions will have to use a range of frequencies not used by any similar type of station within the reach of that broadcast power. On the other hand, if the transmission of energy is in the form of a well-defined beam of energy, any allotted frequency can be used as long as no one has a powerful transmitter in the immediate vicinity with a wide area pattern radiating on the chosen frequency.

Only so much information can be impressed on a given carrier frequency. The limiting parameters depend on the frequency content of the information. In the last chapter, some of the general relationships between the type of modulation and the bandwidth required for transmission of the information were covered. To understand more about the allotment of the communications spectrum it will be worthwhile to preview the requirements of some specific systems that are commonly in use.

Voice Communications

If the purpose of the communication system is to transmit the human voice just to understand it, the bandwidth requirements are relatively low. Most human voice frequencies are lower than 4,000 hertz (4kHz). If these are frequency multiplexed onto a modern coaxial cable, the cable could carry over 10,000 two-way telephone conversations. If the voice transmissions are frequency multiplexed onto a waveguide transmission system of the type shown in *Figure 5-13,* over 200,000 simultaneous conversations could be carried by such a waveguide operating with a bandwidth of 1 billion cycles per second (1 gigahertz).

If digital transmission is used to carry the voice information, the speech signal must be sampled at twice the highest frequency of interest, which would be twice 4 kHz or 8kHz. This can be converted to its digital equivalent by converting the analog speech signal to a 7-bit digital code. This 7-bit code can represent one of 128 possible analog levels, which will be quite accurate enough for intelligible speech transmission. Thus, 8,000 samples will yield 7 binary bits each second for a bit rate of 56,000 bits per second (as mentioned previously, in computer terminology this is 56 Kilobauds). This is the bit rate (considered as frequency) required to transmit human speech with digital techniques.

The cable system used to transmit the digital information must be capable of being pulsed, as a minimum, at this frequency or bit rate. A pair of twisted wires can handle up to 1.5 million bits per second so that 24 voice transmissions could be encoded and time-multiplexed onto such a transmission system. Modern coaxial cable is designed to carry 274 million bits per second which would handle over 4,800 time-multiplexed voice conversations at one time.

If the voice signals are to be transmitted using AM radio techniques, and high fidelity is not important, then an AM system with a minimum bandwidth of 8,000 hertz would satisfy the need. However, to avoid interference with other voice stations, a bandwidth of 50 to 100 kHz is often used for such a transmission. Radiation patterns are not beamed and are quite variable. For this reason, stations are located much further apart in physical location and wider in frequency to make sure no interference occurs. In the standard AM broadcast band, only about twenty such transmissions can occur simultaneously in the same 100 kilometer area.

On the other hand, at the CB frequency range of 26.96 to 27.41 MHz, there are 40 channels allocated for voice communications, over a total bandwidth of 450 kHz. This provides a bandwidth of 11 kHz per communications channel, which is very close to the AM channel requirements of 8 kHz. The CB radios and transmitters achieve this close separation by using very precise and preset tuning methods. If the upper CB carrier frequencies of 72 to 76 MHz are used and the same 11 kHz channel spacing is used, 400 channels can operate simultaneously in the same area.

As the carrier frequency is increased, the 11 kHz channel bandwidth requirements for voice communications becomes less significant as illustrated in *Figure 5-22. Figure 5-22* illustrates graphically the portion of the frequency band required by the 10 kHz voice channel bandwidth when the broadcast carrier frequency is at 100 kHz, 1 MHz and 10 MHz. In the frequency band from 100 kHz to 1 MHz range, there is room for 100 such voice channels. In the band from 1 MHz to 10 MHz, 1000 such voices communications are possible. Each increase in the carrier frequency band by a factor of ten also increases the available number of voice channels by a factor of 10.

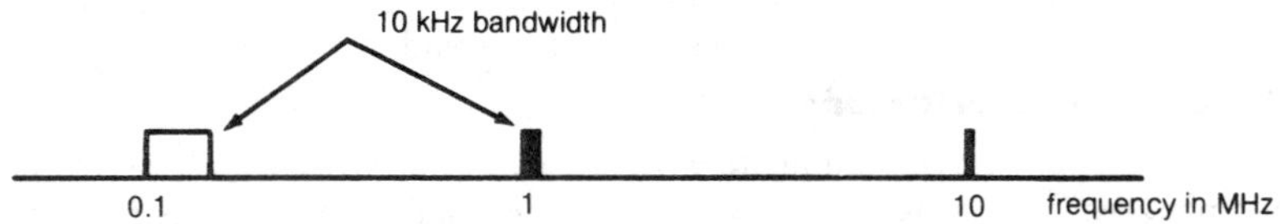

Figure 5-22. *Width of 10 kHz band at 0.1 and 1 MHz Range.*

Television Communications

Color television transmission requires a bandwidth of 4.5 to 4.6 MHz. The carrier (station) frequency must be transmitted at high frequencies in order to provide the room in the spectrum for the information bandwidth required for each channel. This is why the FCC has assigned TV broadcasting the ranges of frequencies in the VHF and UHF portion of the spectrum. Thirteen channels are positioned in the split frequency ranges from 54 MHz to 216 MHz and the additional 56 channels are positioned in the frequency range for 470 to 806 MHz.

If the television picture information is transmitted digitally, approximately 92.5 million bits per second are required. If such program material is transmitted over a coaxial cable with a capacity of 274 million bits per second, only three such programs could be time-multiplexed and simultaneously carried on such a cable.

High Fidelity Music Broadcasting

Stations that broadcast monophonic music require an information bandwidth of 20,000 hertz. Others that broadcast stereophonic music need a bandwidth of 40,000 hertz. If one attempted to broadcast stereophonic high fidelity music on the standard AM broadcast band, each station would require a bandwidth of 80 to 100 kHz. As a result only a few stations could be assigned to the overall broadcast band in a given area.

One further thought about a requirement for high fidelity. The noise susceptibility of standard AM transmission would in many cases, make the fidelity of the music unacceptable. FM transmission is needed to reduce the noise effects, but a large bandwidth FM signal is required to send the information. The stations in the FM broadcast band are positioned with 200 kHz bandwidths in the frequency range from 88 to 108 MHz. Thus, up to 100 stations could operate simultaneously in a given area in this range of frequencies. To attempt to provide such FM high fidelity transmission in the 535 to 1605 kHz range would limit the number of stations to five.

Transmitting stereophonic high fidelity music digitally, using a 7-bit code and sampling at 40 kHz, would require 280,000 bits per second per stereo channel for a total of 560,000 bits per second. If more bits are used to encode the music signal, even higher bit rates would be needed.

Computer Communications

Computers communicate with streams of digital codes. The rate of bits per second involved in the communication usually is limited by the transmission system or by the terminal being used. The central processing hardware is hardly as likely to limit the rate. Card readers and old teletype machines are relatively slow and can send only at the rate of 110 bits per second (called 110 baud in computer jargon). High-speed terminals and printers can send and receive information at much faster rates, with 9600 baud (bits per second) being common. This information is usually transferred in bursts of a few hundred bits at a time, with long periods of inactivity.

Summary

The frequency bandwidth requirements of transmitted analog information and the bit rate requirements of transmitted digital information of some of the systems that will be covered in this book have been presented in a general sense. These requirements will be discussed in more detail in the remaining chapters. Several general observations can be made when viewing these requirements in relationship to the available spectrum. The higher the transmitter carrier frequency, the more bandwidth available to the communications task. This means that more voice, television, computer, or other information can be transmitted using a high-frequency carrier than with a low-frequency carrier or through a high-frequency waveguide than through a low-frequency cable. The more detailed or stringent the information requirements, the higher the bandwidth required. The higher the bandwidth required, the higher the bit rate required in a digital transmission. The use of frequency and time-multiplexing allows a given transmitter-receiver system operating at a given frequency to carry the maximum number of information channels.

As carrier frequency is increased, dimensions for antennas and other hardware are decreased; transmission through air becomes line-of-sight; atmospheric conditions of rain, sleet, and snow degrade the

transmission; and transmission must be reinforced by repeated amplification. All of these factors must be considered for each type of communication system and the information it is to carry if the available spectrum is to be used efficiently.

WHAT HAVE WE LEARNED?

- The electromagnetic spectrum comprises the range of frequencies from a few hertz to above the frequencies of visible light.
- Very low frequencies are not used in modern communications systems.
- Electromagnetic radiation in the frequency range of 0.5 to approximately 30 MHz are reflected from the ionosphere at the top of the earth's atmosphere so that long distance communications become possible.
- Electromagnetic radiation above 100 MHz travels in a straight line from the transmitter to the receiver. Transmission in this range of frequencies occurs over relatively short distances. Long distance transmission usually requires a reception and retransmission by relay stations.
- Electromagnetic radiation in the microwave range (frequencies above 1 GHZ) can be directed into narrow beams by relatively small antennas. This allows several communications systems to use the same frequencies in the same area without interfering with each other.
- Cables and waveguides can be used to completely control the path of the electrical or electromagnetic energy used in a communication system, totally eliminating the effect of interference from other nearby communications systems.
- The use of the spectrum by communications systems using electromagnetic radiation is regulated in the United States by the Federal Communications Commission or FCC.
- Both national and international regulations allow citizens to use the electromagnetic spectrum for their individual communications or experimentation through the amateur radio bands.
- Different types of information require different information bandwidths and different digital transmission bit rates. The carrier frequency and type of transmission used in the communications system must take these requirements into account.

WHAT'S NEXT?

Now that the basic concepts of communication, electronic communication, and communications spectrum have been examined, it is possible to start learning about specific types of communications systems. The rest of this book will deal with the most common types of communications systems in use today, beginning with the telephone and telegraph systems in the next chapter.

Quiz for Chapter 5

1. A sinusoid of 300 MHz that is travelling at the speed of light in the form of electromagnetic radiation has a wavelength of:
   a. 0.1 meter.
   b. 1 meter.
   c. 10 meters.
2. A device that emits visible light is:
   a. generating electromagnetic radiation.
   b. generating light waves whose wavelengths is in the millionths of meters.
   c. is producing energy whose sinusoidal frequency is much higher than that of radio waves.
   d. All of the above.
   e. None of the above.
3. The ionosphere:
   a. is a layer of electrically charged particles.
   b. acts as a reflector for radio waves in the microwave range of wavelengths.
   c. is used to provide long distance communications with medium to high frequency radio waves.
   d. All of the above.
   e. a and c above.
4. Radio waves in the VHF range up to 10 GHz (billion hertz) range:
   a. are not reflected by ionospheric or atmospheric effects.
   b. are very susceptible to fading.
   c. are predominantly line-of-sight radiation.
   d. All of the above.
   e. a and c above.
5. If VHF and higher frequency radiation is to be focused into a beam by an antenna, the antenna diameter required for a given beam width:
   a. increases with frequency.
   b. decreases with frequency.
   c. is independent of frequency.
   d. None of the above.
6. Rain and snow attenuate radiation for frequencies:
   a. above 10 MHz.
   b. above 1 GHz.
   c. above 10 GHz.
   d. below 10 GHz.
7. A repeater station:
   a. receives radio transmissions and transmits this information to another repeater or receiver station.
   b. Only receives information.
   c. Relays line-of-sight radiation from previous stations to the next station so that communications over distances greater than line-of-sight limitations can be achieved.
   d. All of the above.
   e. a and c above.
8. Communications cables:
   a. Allow many users in an area to share the same portions of the spectrum by constraining the energy along a metallic path instead of allowing it to radiate in all directions through space.
   b. are very susceptible to noise influences from the surrounding environment.
   c. are limited in useful frequencies to below 10 MHz.
   d. All of the above.
9. The use of the radiated electromagnetic spectrum is regulated by:
   a. international treaty.
   b. the Federal Communications Commission in the United States.
   c. is left to State and local governments.
   d. individual agreements between transmitter stations.
   e. a and b above.
   f. a and d above.
10. Which broadcast system has the highest bandwidth requirement?
   a. AM commercial broadcast station.
   b. FM commercial broadcast station.
   c. CB radio station.
   d. voice amateur radio station.

(Answers in back of the book)

Telegraph and Telephone Systems

ABOUT THIS CHAPTER

In the previous chapters the general nature of electronic communications and the transmission of electronic information were covered. In this chapter a specific type of communications system will be considered. This system is the modern telegraph and telephone system that exists throughout the world to provide a relatively complete communications capability for transferring information from one point to another. It is a very familiar system since telephone communications is a commonplace activity in everyone's life. It has all the features required of a good communications system in that it is simple to use, effective, and reliable, so much so that it is often taken completely for granted. The present system has evolved over many decades and is supported by a very large, and in some cases very complex, electronic technology. In this chapter, the basic concepts behind the system and its technology will be examined in detail.

WHAT ARE THE BASIC FEATURES OF TELEGRAPH AND TELEPHONE SYSTEMS?

The Telegraph System

The early telegraph and telephone systems were simply a wired connection between the two communicating points. As discussed briefly before, in the early telegraph system (*Figure 6-1*) the transmitter was a spring action switch held for the correct time that closed or opened a circuit to apply or interrupt a voltage to the telegraph line. At the receiving location the resulting line current caused a mechanical sounder to click out the short and long pulses sent from the transmitter. The information in the telegraph system is digital in nature, consisting of a code of long and short duration pulses of line current.

As shown in *Figure 6-1,* each end of the telegraph line has a transmitter and receiver. Normally the send-receive switch at each end is on the receive position so that any sending by a transmitter will be received. Only one transmitter can be on the line at any one time. If a transmission is being received no other transmitter should be connected to the line. (Actually, the open telegraph key keeps the transmitter off the line, therefore, strictly speaking the send-receive switch would not be required in this simple example.) Note that the send-receiver station in the middle of the line (Station C) also can receive and send signals.

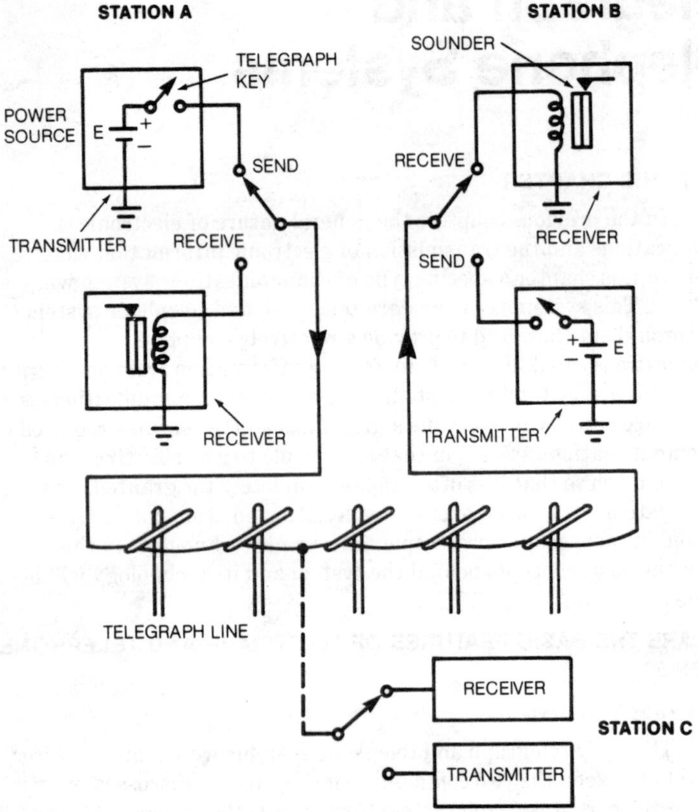

Figure 6-1. *Basic Telegraph System.*

The primary difference between telegraph and telephone systems is in the types of information handled by the systems, which in turn depends on the types of transmitters and receivers used. In the telephone system of *Figure 6-2* the telegraph key and sounder at each end of the line is replaced by the telephone handset. The handset contains a speaker and a microphone which are the receiver and transmitter, respectively. As shown in *Figure 6-1,* the telegraph system is a two-way system but of a restricted nature. The telephone system is a complete two-way system (the technical term is full-duplex), transmitting and receiving can occur at the same time from both connected stations.

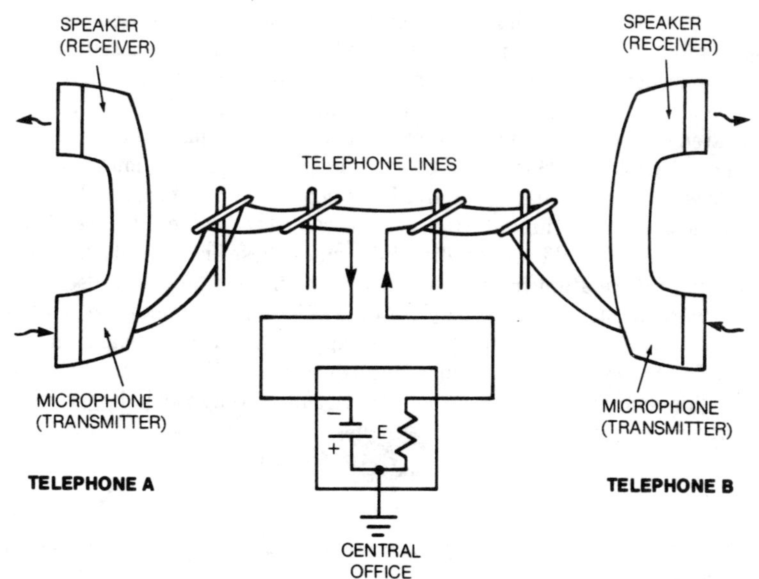

Figure 6-2. Basic Telephone System.

The Telephone System

In the telephone system, the two-way communication is carried by either a two-wire or a four-wire system. In the four-wire system one pair of wires is connected from the transmitter at telephone A to the receiver at telephone B, and the other pair of wires is connected from the transmitter at telephone B to the receiver at telephone A. In the two-wire system, which uses one-half the wiring of the four-wire system and therefore is less costly, both transmitter and receivers at telephones A and B share the same pair of wires for the two-way communication. The power for the communication is provided by a battery (or an appropriate power source with battery backup). The battery is usually provided at the central office in commercial telephone networks, though it can be provided at the transmitter in a private or special purpose four-wire system.

The sketches in *Figures 6-1* and *6-2* are very limited systems. Only a very small network (circuit) is shown, and just two or three stations. In a modern system many multiple paths are available to allow connection of any given telephone to another selected from the hundreds of millions of telephones available throughout the world. In addition, in *Figures 6-1* and *6-2* the wires are shown connected from one point to another through the use of open wire strung on telephone poles. In modern telephone systems many ways exist for transmitting the information from one point to another.

The options are shown in *Figure 6-3*. Which option is used for a given communication depends largely on where A and B are located. If A and B are located close to each other, the information will be carried over wires on telephone poles or in overhead or underground cables. If A and B are located on the same land mass but a large distance apart, the information could be transmitted over telephone lines, cables, microwave links, or even satellite links. If A and B are separated by an ocean, the transmission must be through transoceanic cable (yes, cables have been laid on the ocean floor), radio, or satellite links, since these are the only feasible ways to get information from one continent to another. Most of these options were discussed in the last chapter. The satellite communications approach will be covered later. Whichever approach is used, a means must be provided for selecting the proper paths over which the information is to travel. This requires some way to connect circuits through switches called network switching.

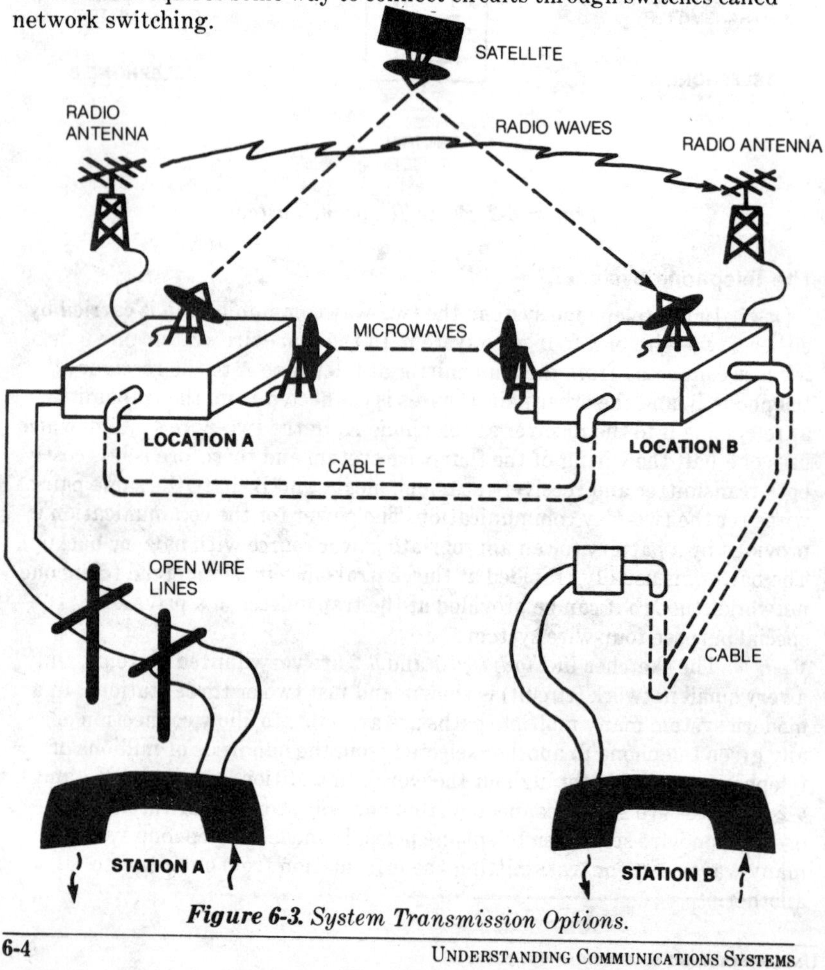

Figure 6-3. *System Transmission Options.*

Network Switching

The basic requirements for network switching are summarized in *Figure 6-4*. Each telephone in the system must be assigned a number to indicate its location in the system. The switching network must recognize which telephone is initiating the call and which telephone is to receive the call. From this information, it must set up the circuit connections for a signal path that will send the information from the sending (calling) telephone to the receiving (called) telephone. In commercial telephone systems a seven digit number is sufficient to locate individual telephones within a metropolitan area. As shown in *Figure 6-4*, digits specify which central office supplies the power to the telephones that are to be connected and the last four digits determine which telephone of a possible 10,000 telephones is calling or is being called. The central office recognizes which telephone is calling (initiating the call) when the telephone is removed from its hook or cradle. It recognizes which telephone is called by the number that is dialed by the calling telephone. If the call is to a telephone outside the metropolitan area, an additional three digits are used to define the area code or location of the called metropolitan area. This would then be a long distance call.

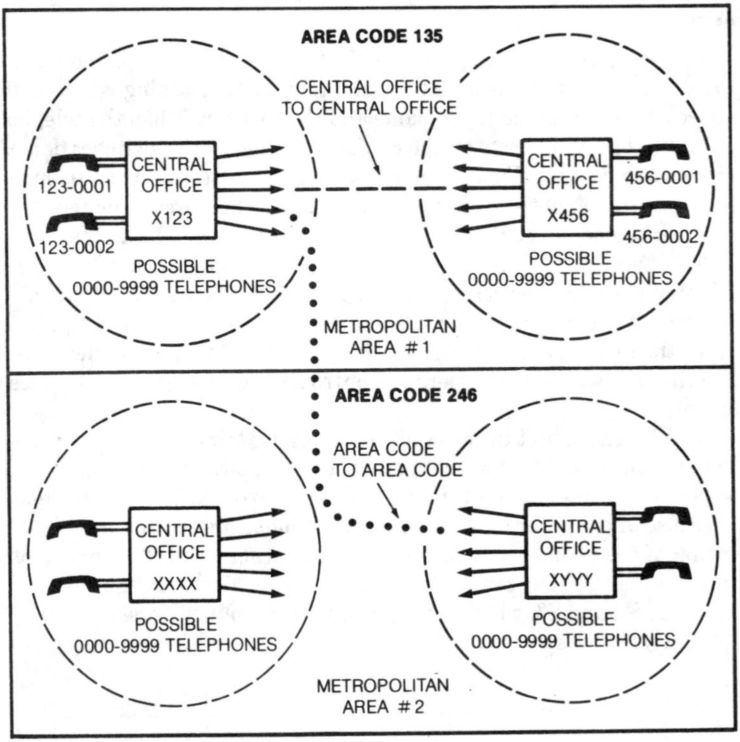

***Figure 6-4.** Basic Telephone Network.*

Figure 6-5 shows a very simplified view of the switching needed to allow a telephone connected to central office A to call another telephone connected to central office A or one of the telephones connected to central office B. If central office A has an exchange code of 123 and central office B has an exchange code of 456, then a call from telephone number 123-0001 to number 123-0003 would require a circuit connected by the switched path as indicated by the dashed line in *Figure 6-5*. If on the other hand the telephone at 123-0001 is calling the telephone number 456-0002, the switching network would have to set up the switched circuit path indicated by the dotted line. Of course, the network has to do much more than what's shown in *Figure 6-5*. It must provide the switching circuitry allowing for the possibility of many calls occurring at the same time within a central office and between central offices. Thus, many parallel switch paths must be available. Furthermore, the network also provides detection circuits for all of the signalling which determines the calling and called telephones and all of the line checking required to establish that the lines and telephones are clear to complete the communication. These signalling requirements depend on the characteristics of each telephone set and on the signalling sequence required to make a telephone call.

Handset

The parts of the telephone handset are shown in simplified form in *Figure 6-6*. The instrument used for talking into and listening will be called the subset. It contains the microphone and the speaker. When the telephone is not in use the subset rests on the cradle which opens the switches denoted SH (for switch hook) in *Figure 6-6*. These switches disconnect the telephone subset from the telephone system. However, there is a circuit connection that is maintained to the handset. An electromagnet called the ringer solenoid is connected to the telephone line wires on the central office side of the SH switch so that the central office can ring the telephone with an ac signal when it is called. The telephone wires are denoted as the T and R lines, for the terms Tip and Ring, which were related to plug connections used in the original manual (operator controlled) switching central offices or exchanges.

With the subset in the cradle (or hanging telephone may be on a hook – thus the name switch hook) no dc current flows from the central office 48-volt talking battery through the T-R loop or thru the ring solenoid loop because of capacitor C. The central office monitors the dc current to determine if the phone is idle or busy or is initiating a call. The central office controls the signals that can be delivered to the handset through the switches S1, S2, and S3, which are again a very simplified version of the actual central office switching control center.

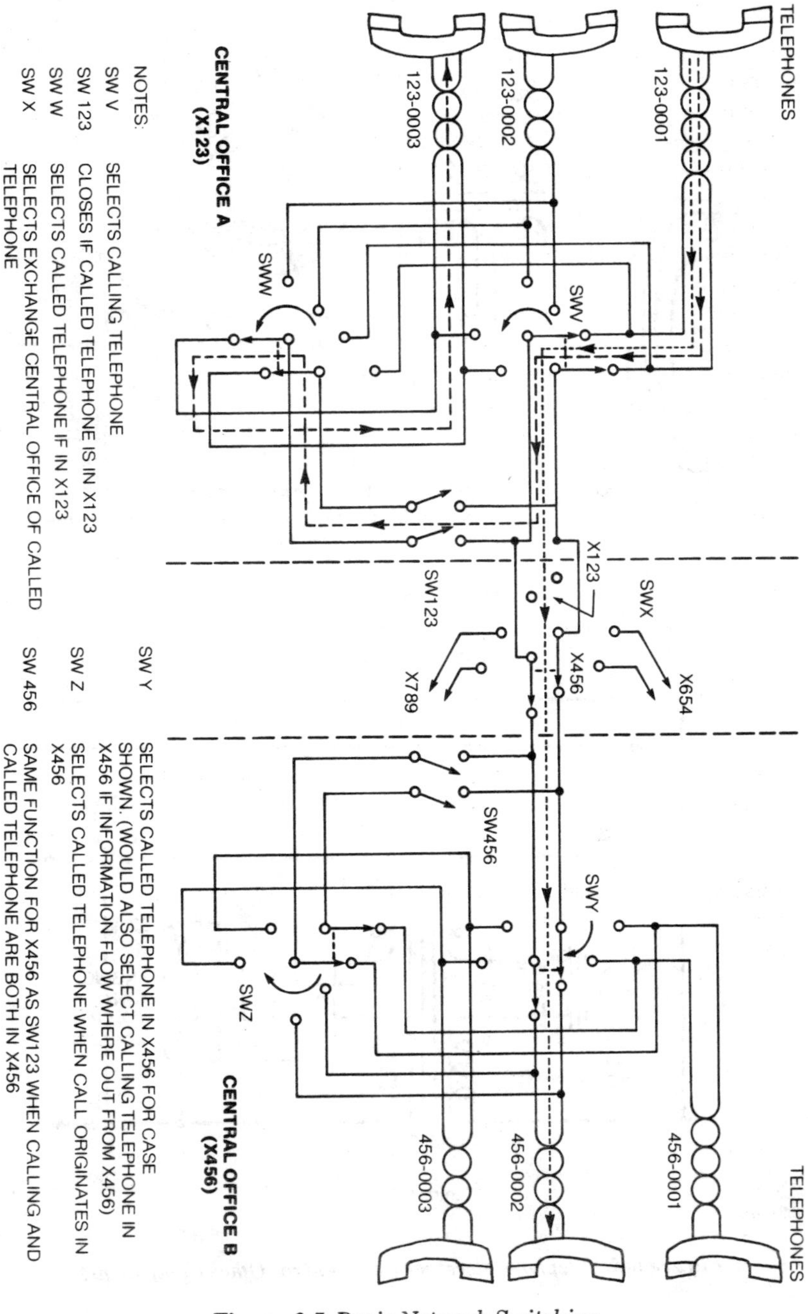

TELEPHONES

123-0001

123-0002

123-0003

SWV

SWW

CENTRAL OFFICE A (X123)

SW123

X123

SWX

X654

X456

X789

SW456

SWY

SWZ

CENTRAL OFFICE B (X456)

456-0001

456-0002

456-0003

TELEPHONES

NOTES:

SW V SELECTS CALLING TELEPHONE

SW 123 CLOSES IF CALLED TELEPHONE IS IN X123

SW W SELECTS CALLED TELEPHONE IF IN X123

SW X SELECTS EXCHANGE CENTRAL OFFICE OF CALLED TELEPHONE

SW Y SELECTS CALLED TELEPHONE IN X456 FOR CASE SHOWN. (WOULD ALSO SELECT CALLING TELEPHONE IN X456 IF INFORMATION FLOW WHERE OUT FROM X456)

SW Z SELECTS CALLED TELEPHONE WHEN CALL ORIGINATES IN X456

SW 456 SAME FUNCTION FOR X456 AS SW123 WHEN CALLING AND CALLED TELEPHONE ARE BOTH IN X456

Figure 6-5. *Basic Network Switching.*

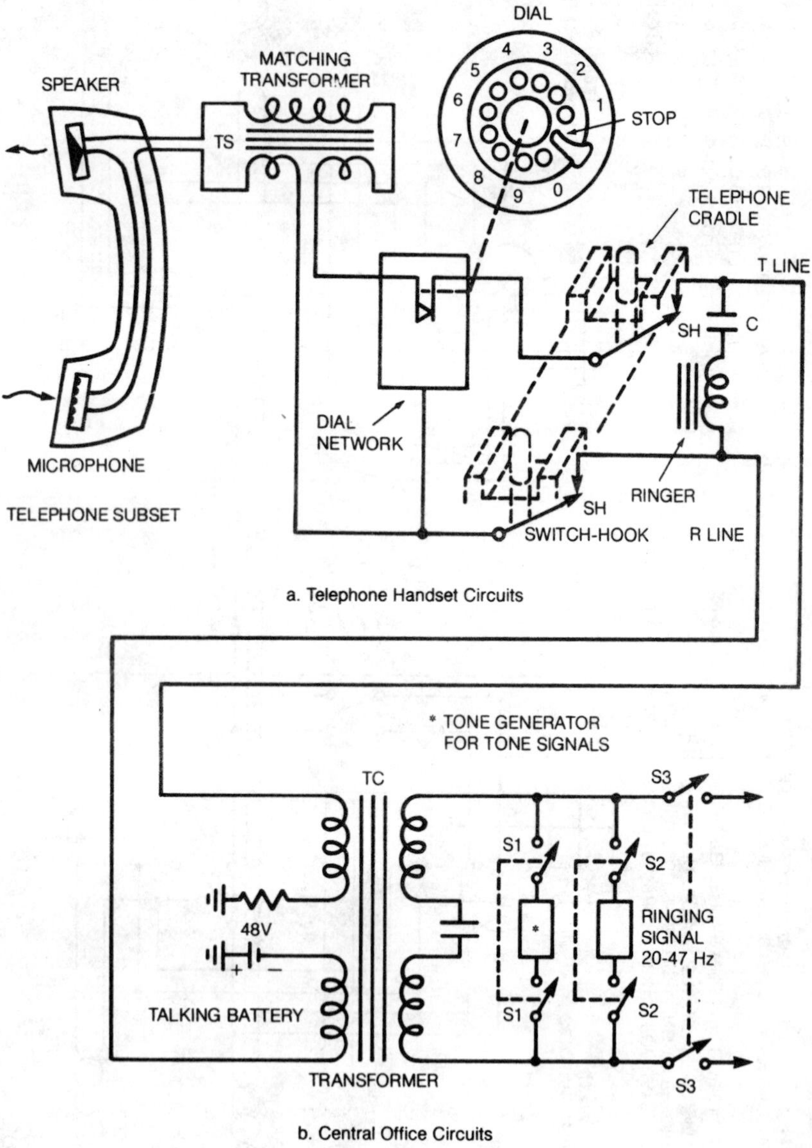

a. Telephone Handset Circuits

b. Central Office Circuits

Figure 6-6. *Telephone Receiver and Central Office Components.*

Ringing A Called Telephone

The signals that can be sent to the handset are summarized in *Figure 6-7*. These signals are familiar to all telephone users, as is the sequence in which they are used. The sequence will be explained using the terms of *Figure 6-7* and the circuitry of *Figure 6-6*. Two cases will be considered. First, assume that the telephone in *Figure 6-6* is idle – the subset is on the cradle or on the hook. It is ready and waiting to receive an incoming call as indicated by the lack of dc current flow in its T and R lines. If this telephone has been selected by the central office to receive a call, S2 will be thrown to connect the ringing signal to the T and R lines through the transformer TC. The 110 volt ac signal of 20 to 47 hertz will cause the ringer solenoid to sound the familiar telephone ring at the handset in *Figure 6-6*. This ringing will continue until the caller hangs up or until the telephone in *Figure 6-6* is answered by lifting the subset off the switch hook. Such action will cause dc current to flow in the T-R loop through the subset. The central office circuitry will detect the dc current flow and remove the ringing signal by opening S2. It also will throw switch S3 to connect the set to the calling transmission path.

Signal Tone	Interrupt Rate	Frequencies (Hertz)
Dial	None	480
Ringing	None	440 modulated with 480
Busy	60 (Called Line)	480 modulated with 620
	30 (Toll Line)	
	120 (Intra-office)	

Figure 6-7. *Basic Signalling Tones in Telephone System.*

Answering The Telephone

As the speaker talks into the subset of *Figure 6-6* to answer the phone, the microphone of the handset causes the current in the T-R loop to vary to produce electrical signals that correspond to the pattern of the speech waveform. The T-R loop current through the matching transformer TS and the primary of the transformer TC produces the speech signals in the secondary circuit of the transformer TC. The secondary of the transformer TC is connected to the calling telephone through the circuit path formed by the closing of switch S3 and network switching circuits in the central office similar to those discussed in *Figure 6-5*.

Now if the handset of *Figure 6-6* is considered to be the calling handset, after the person calling has dialed and the called telephone owner has lifted the subset and spoken into it to answer the telephone (as discussed above), the electrical signals representing the spoken word are now the input signal to the transformer TC through the closed switch S3. The S3 side of the transformer is now the primary and the T-R loop side is the secondary. The current in the primary produced by the speech signals induce a current in the secondary which excites the speaker of the handset through the matching transformer TS and reproduces the spoken word of the person answering the called telephone in the ear of the person at the calling telephone. As each person speaks at either the called or calling telephone, the spoken word is reproduced from transmitter to receiver through the completed communications circuit. The system reproduces the spoken word in both directions through the same circuit.

One other point about the handset of *Figure 6-6*. Whenever a person speaks into the microphone of the handset they also hear their spoken words in the speaker of the same handset. The matching transformer TS determines the correct signal level for this signal as well as the correct signal level for the signal received from the sending telephone.

During the above described sequence the dialer sub-circuit is not used, the tone signalling circuit is not used, and switch S1 remains open.

Initiating A Call

The second case is when the handset of *Figure 6-6* is used to initiate a call. It is the calling telephone. Both the dial network and the tone signal generators are needed to provide the proper basic signalling tones of *Figure 6-7* between the calling telephone and the central office. When the caller lifts the subset off the cradle (or hook for a hanging subset) dc current flows in the T-R loop. This is sensed by the central office and indicates that a call has been initiated. The central office locates the line initiating the call and places a selected tone signal generator output of 480 hertz on the T-R line by closing switch S1 and producing a current in the primary of transformer TC. This produces a dial tone in the calling handset's speaker. Switches S2 and S3 are open (S2 will remain open, since no ringing current will be delivered to the calling telephone). When the caller hears the dial tone, the number of the telephone to be called will be indicated as an electrical signal by a rotary dial containing switch contacts that open and close or by pressing touch-tone pushbuttons. The dial tone will be interrupted when the first digit is dialed with a mechanical dialer. In the case of the touch-tone system it will be replaced with the tone of the button pushed. These tones are summarized in *Figure 6-9* and will be referred to later in this discussion.

Mechanical Dialer

If a mechanical dialer (*Figure 6-6*) is used, the dial will be rotated according to the digit required and allowed to rotate back to its rest position. As it does it will interrupt the dc current in the T-R lines to cause a number of pulses equal to the number dialed. The dial is mechanically designed to provide 10 pulses per second. The speaker is usually disconnected by the dial network to keep from sounding the dialing clicks. The central office circuits receive the pulses and detect the sequence of numbers dialed and stores them. They use these numbers to locate the called telephone and to set up an available transmission path between the two stations. When the called telephone is located, the central office checks the on-hook status of the called telephone, and rings it if the telephone is on-hook (idle). It also places a ringing tone through the closed S1 switches (S2 and S3 are still open) on the line so it is heard at the calling telephone. This ringing tone is developed by modulating a 440-hertz sinusoid with a 480-hertz sinusoid. This results in a 920-hertz tone increased and decreased in intensity at a 40-hertz rate as shown in the pattern of *Figure 6-8*. The ringing tone continues until either the called telephone answers or the calling telephone hangs up. If the called telephone answers, the central office circuits remove the ringing current from the called set, the ringing tone from the calling set and completes the transmission path by closing the switches S3.

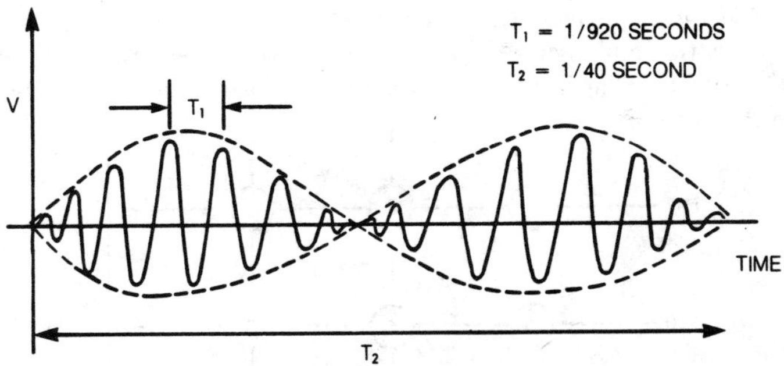

Figure 6-8. *Ringing Tone Waveform.*

Busy Telephone

If the called telephone is busy (off-hook) the tone generator will sound one of three possible busy signals to the calling telephone (*Figure 6-6* with switches S2 and S3 open). The busy tone is a 140-hertz tone generated by modulating a 480-hertz sinusoid with a 620-hertz sinusoid and filtering out the 1100-hertz signal (when one sinewave modulates another, both the sum and difference frequency sine wave signals result). This tone is interrupted at a rate that indicates the reason for the busy signal (*Figure 6-7*). A signal interrupted at the rate of 60 times per minute indicates the called line was busy. If the interrupt rate is 1/2 of this, the toll line between the central offices was busy (full of existing calls). If the interrupt rate is doubled to 120 times per minute it implies that all intra-office paths are busy. Only with the 60 interrupts/minute is the caller sure that the called party line is busy.

Touch-Tone Dialing

When the handset of *Figure 6-6* is a touch-tone telephone, then the dial network is more than just switches. It contains tone generating circuits as well. When the touch-tone telephone is used, the touch-tone key or pad causes a signal of two frequencies to be placed on the line. The frequencies are indicated by the intersection of the frequency lines in the touch-tone matrix of *Figure 6-9*. For example, pressing the 5 key causes a 770-hertz and a 1336-hertz tone to be sent to the central office (and to the called party if the button is pushed during a conversation). Central office circuits that detect and decode the tones set up the switching for the communications path just as for the mechanical dialer. The use of such tones speeds up the dialing operation and allows command and control information to be sent to the called location. The latter use of the touch-tone signals will be discussed later in the chapter.

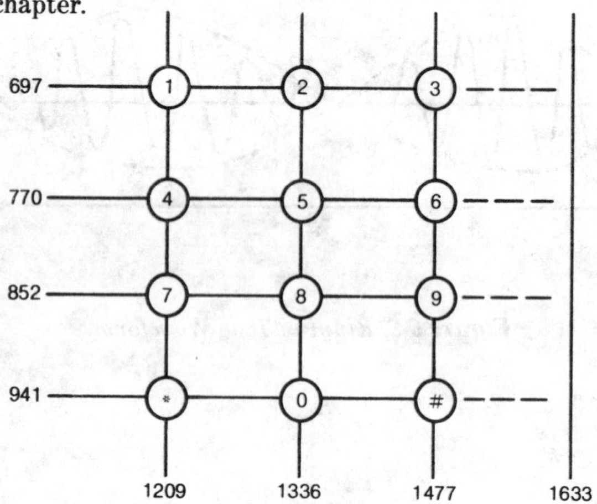

Figure 6-9. *Touch-Tone Frequency Assignments.*

Bandwidth

The frequencies of the tones generated by the touch-tone keys as well as the conversations of the telephone users must be within the bandpass capabilities of the telephone system. The bandwidth of a channel in a typical telephone system used for conversations is 4 kilohertz as shown in *Figure 6-10.* The bandpass region allowed for the voice signals is 200 hertz to 3400 hertz. The touch-tone signals fit into this range as do some special control signals at 2400 and 2600 hertz. Other control signals from 3400 to 3700 hertz fit into the overall 4 kilohertz bandwidth of the telephone channel. Also shown in the dashed line curve of *Figure 6-10* is the energy distribution versus frequency of typical human speech. Combining this with the 200-hertz to 3400-hertz bandpass for the voice signals shows that the system can provide good understanding of the words spoken and good identification of the speaker. This voice channel bandwidth will be an important system parameter throughout this chapter. The other important parameters of the voice channel are the signalling tones and the dc currents that are used by the switching and control circuits of the central offices to establish the communications paths between many pairs of telephones over a worldwide network. Understanding the basic concepts of the central office circuits will help to further understand the telegraph and telephone communications systems.

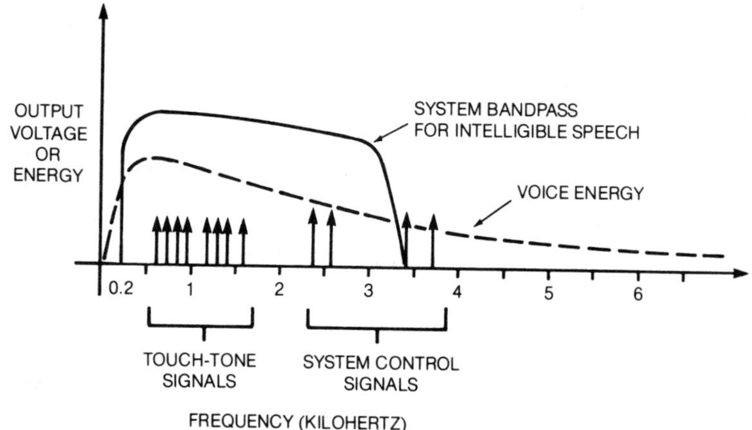

Figure 6-10. *Voice Channel Bandpass and Signal Features.*

HOW ARE SIGNALS SWITCHED IN TELEPHONE NETWORKS?

The basic types of switching requirements were presented in *Figure 6-4* and *6-5*. Previous discussion showed how telephones that have the same first three digits in their telephone number must have their communications paths interconnected in the same central office to complete these intraoffice conversations. Telephones with different first three digits must have the calling telephone's central office connected to the called number's central office before a final transmission path can be established. If the central offices are in different parts of the country, the switching can be much more involved than the case in *Figure 6-5*, since area codes and world codes may be involved in detecting and decoding the location of the called telephone in the system. A part of this more extensive network is shown in *Figure 6-11*.

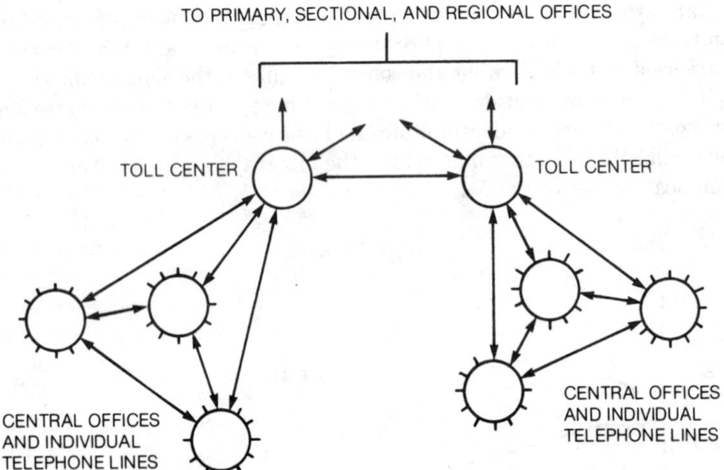

Figure 6-11. *Portion of National Telephone System.*

If a call is placed from one area code to another, the call will have to at least pass through a toll center in each area. The calling toll center must search for an available connection to the called toll center, which will in turn determine a path to the correct central office which will then find a connection to the called telephone. If the calling toll center cannot find a direct path to the called toll center, the call will be routed to higher levels (called primary, sectional, and even regional centers) each of which searches for a path from the calling to the called toll center. At each level, the search for an idle path involves either examining for dc current flow in the path or for sinusoidal currents of a given frequency in the path.

When the path involves a radio, microwave, or satellite link, completing the communications path becomes even more involved. A computer search is usually required. The messages are usually given a priority by central computers keeping track of which paths are busy. Based on the type of communication and its priority, they will select the most efficient path for a given communication. Whatever the approach, for local calls, especially in some of the early central offices using electromechanical switching techniques, the most common approach to detecting an idle path is to detect a dc current.

DC Current Detection

The basic idea behind dc current detection using relay switching is shown in *Figure 6-12*. *Figure 6-12a* shows a two-wire transmission line and the basic interconnection along the transmission path. S1 and S2 are contacts on relays. *Figure 6-12b* indicates the current flow for the possible combinations of S1 and S2. The current can be sent through relay coils to control the state of the system at both ends of the communications path. In fact, if both telephones are in the same central office, the switches S1 and S2 are relays whose position is controlled by currents flowing in the telephones of the calling party and the called party. If the telephones are in separate central offices, the currents would be determined by S1 and S2 contacts on relays designed to examine and select available communication paths in the central offices which have been called. The examining and selecting can be understood better by considering a normal sequence of events.

Initially the trunk is idle and no trunk request has been received. Thus, S1 is in position 2, which grounds the A end of the trunk and S2 is also in position 2, which grounds the B end of the trunk. No current flows in the trunk, and system relays would show the trunk available. When a trunk is requested by the system throwing S1 to position 1, current will flow down the trunk in the direction shown in *Figure 6-12a*, assuming the line is still idle (S2 still in position 2). This current is used to close a set of relay contacts at point A granting station A the use of the trunk to ring the called telephone. When the called party answers, the currents from station B to the called line will throw switch S2 to position 1. The dc currents will go to zero, indicating the line is now busy. Other stations wanting to use the line can detect this busy status by the fact that S1 and S2 are in position 1 (either can be checked) and that the line dc current is zero, which is indicated by the state of various system relays. Only if S1 or S2 are in position 2 (ground on a given end of the line) and no line current is flowing is a line available.

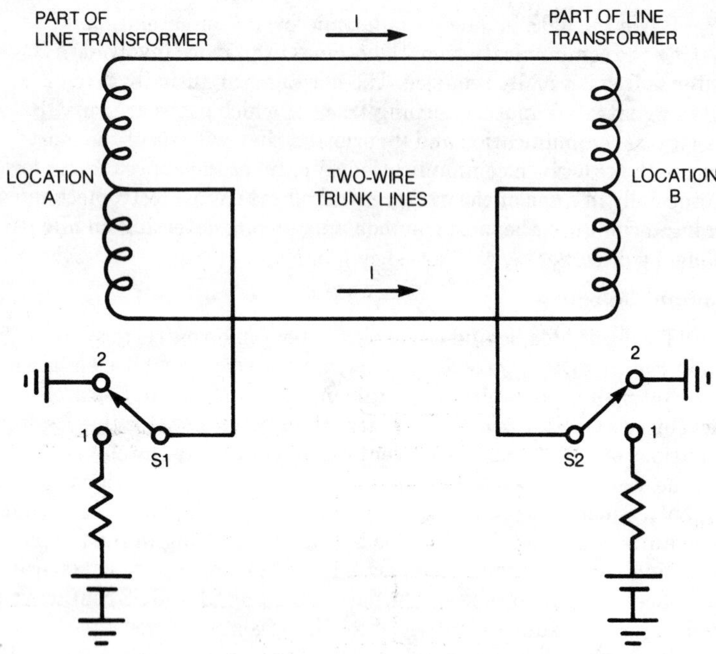

a. Two-Wire Interconnections

SWITCH POSITIONS			
S1	S2	I	TRUNK CONDITION
2	2	0	Idle
1	2	+I	Seized by Caller
2	1	–I	Unavailable
1	1	0	Busy-Call Connection Made

b. Current Flow For Different Interconnections

Figure 6-12. *Selecting Lines on Basis of DC Current.*

When either party hangs up, the switch at that end of the line will go back to position 2 causing current to flow indicating the disconnect condition. When both parties have hung up, both switches are in the position 2 and the line current is zero, indicating an available line. Of course, ac currents could be used to indicate such a status. In fact, newer offices that utilize touch-tone telephones use the presence of a tone on the line to indicate line availability and the removal of such a tone to indicate line busy. Two tones are used in two-wire systems, one when the signal flow is A to B and the other for signal flow from B to A. One tone is used in four-wire systems.

Interconnection Detail

Once the system has a method for determining if a given line is available, it can provide connections to lines in one of several ways. Electromechanical switching centers use either a sequential switching approach in Step-by-Step switching systems (called SxS system) or a matrix switching approach in Crossbar switching systems. Electronic switching centers use a variety of space, time, and frequency switching under computer control to achieve system interconnections. The basic components of the SxS system are shown in *Figure 6-13*.

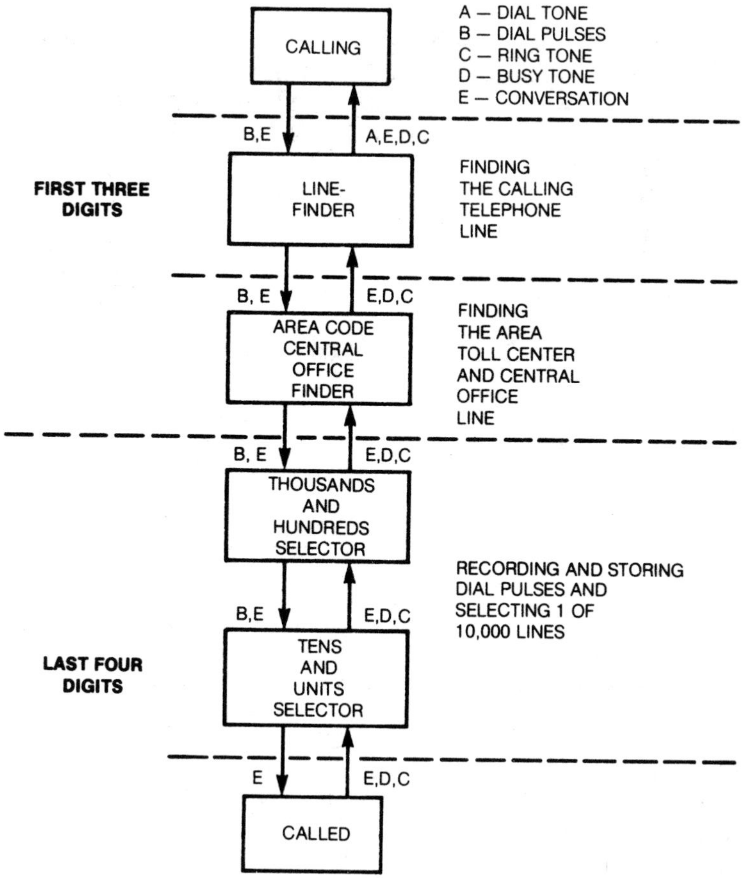

CALLING
A — DIAL TONE
B — DIAL PULSES
C — RING TONE
D — BUSY TONE
E — CONVERSATION
B,E
A,E,D,C
FIRST THREE DIGITS
LINE-FINDER
FINDING THE CALLING TELEPHONE LINE
B, E
E,D,C
AREA CODE CENTRAL OFFICE FINDER
FINDING THE AREA TOLL CENTER AND CENTRAL OFFICE LINE
B, E
E,D,C
THOUSANDS AND HUNDREDS SELECTOR
RECORDING AND STORING DIAL PULSES AND SELECTING 1 OF 10,000 LINES
B,E
E,D,C
LAST FOUR DIGITS
TENS AND UNITS SELECTOR
E
E,D,C
CALLED

a. Complete Selection Block Diagram

Figure 6-13. *Step-by-Step Switching System.*

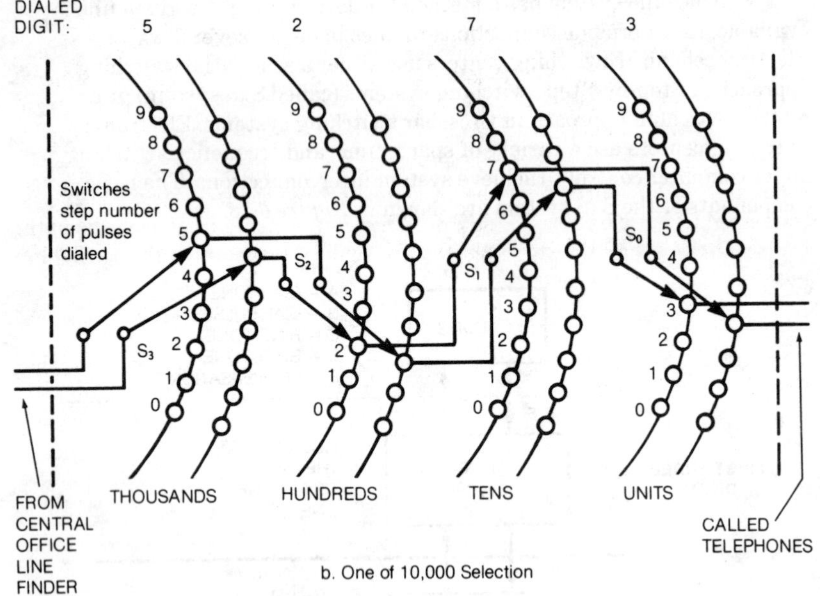

Figure 6-13. *Step-by-Step Switching System.*

SxS System

When the calling telephone is lifted off-hook, the central office searches for the calling line with a linefinder relay rack (performs same function as switch V in *Figure 6-5*). When the line is located, the central office circuits send a dial tone to the calling telephone, which in turn sends its dial pulses to the central office. As shown in *Figure 6-13a,* the dial pulses for the first three digits of the standard seven-digit number indicate whether the called line is in the same central office or in a remote central office. In addition, if the called phone is outside the major metropolitan area, a three-digit area code must be dialed before the standard seven-digit number. If a remote central office is called, an available outgoing trunk is selected and the information routed to the remote office. In the remote central office or in the same central office, the stored last four digits dialed select the called telephone. As shown in *Figure 6-13b,* the thousands digit causes a ten-position stepping switch to rotate to the correct line. This line selection reduces the remaining telephones to be selected to 1,000. Similarly

the hundreds digit causes a second stepping switch to rotate to the correct line to reduce the remaining telephones to be selected to 100. The last two digits cause a similar stepping switch connection to reduce the telephones selected to 10 and then to the called phone. When the final line is connected the called telephone line is checked to see if it is busy. If it is, a busy tone will be returned to the calling telephone. If it is idle, ringing current will be applied to the called telephone and a ringing tone will be sent to the calling telephone. When the called telephone answers (lifts off-hook), the ringing tones and ringing current will be disconnected and the two parties are connected to carry on a conversation. When either party hangs up, the path is disconnected, and the stepping switches will be returned to the available pool.

Crossbar System

The other major type of electromechanical switching system is the Crossbar System, illustrated in *Figure 6-14*. This approach uses a set of matrix relay switches that can quickly set up a transmission path to either a remote central office (inter-office) or to the same central office (intra-office). If telephone A is calling telephone B in the same central office, the central office determines that telephone A has lifted its subset to place a call. It then returns a dial tone to A. Upon receipt of the first three digits of the dialed number, the central office decodes the digits as an intra-office call, the control circuits switch line A to an intra-office line through the line-to-line link frame and the line-to-trunk link frame. The next four digits then define the final connection in the line-to-line link frame to complete the path to B. The interconnections are made similar to that shown in *Figure 6-15*. If the lines Y1 and X7 are to be connected, relay line D would have to receive current to energize the relay and close the contacts. Normal busy and ringing sequences occur as usual along this path.

Suppose C is now the calling telephone and the first three digits of the dialed number indicate an inter-office call to telephone D, the line-to-line link frame makes the connection to the line-to-trunk link frame from the calling line C. The control will select an available trunk line and establish this connection in the line-to-trunk link frame. The last four digits will be sent to a control register in the remote central office, which similarly will set up the switches in its line-to-trunk link and line-to-line link frames to provide an available path to the called line D. The banks of marker relays keep track of which path is used and which paths are available in each central office.

In order for central offices with different types of switching systems to communicate with one another, some interface circuits are required. These are needed to convert tone dialing to pulse dialing and vice versa. Further conversions are required because modern central offices are converting to electronic switching. These systems will require additional interface circuits.

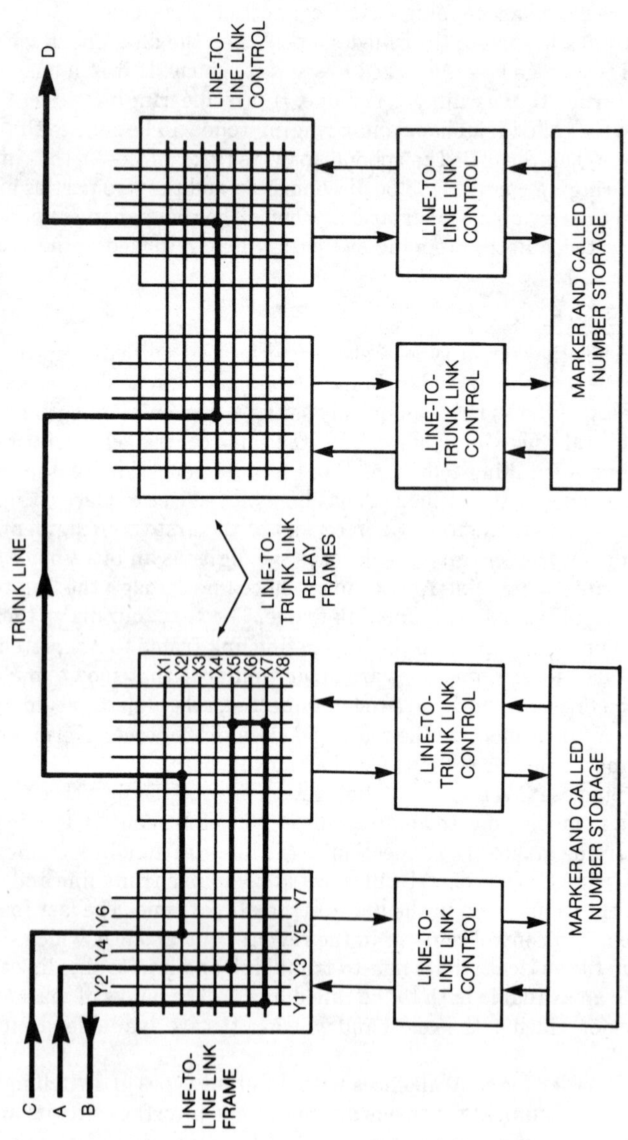

Figure 6-14. *Crossbar Switching System.*

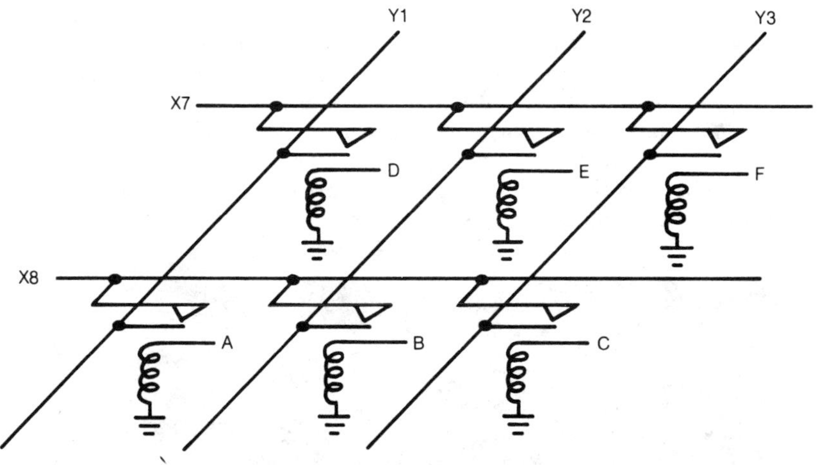

Figure 6-15. *Matrix Connections.*

HOW IS SWITCHING AND MULTIPLEXING USED IN ELECTRONIC SWITCHING SYSTEMS?

Electronic switching offers several advantages over electromechanical switching. Some of these advantages are related to system costs and others are related to system performance. Cost will be considered first. Electronic switching circuits have advantages in increased reliability, lower power, smaller size, lower cost per system function, and standardized functions. Each of these contribute significantly to lower system cost. Reliability means longer system life and lower maintenance costs. Lower power means less energy to be purchased to power the system. Smaller size means significant saving in the space used to house the systems thus reducing capital investments for land and buildings. Standardized functions and lower cost per function means that system design cost will be lower and system material cost will be lower, not to mention the fact that new system designs will come into use earlier in the marketplace making the cost savings available sooner.

In addition to these cost savings, system performance is not sacrificed. In fact, with electronic systems, performance is expanded, even at a lower cost than before. For example, *Figure 6-16* shows a PABX telephone system. This is a private automatic branch exchange system used to serve individual businesses or governmental agencies. This system replaces an entire room of mechanical switching equipment required to do the job, and also provides many optional features that were not available before. Automatic forwarding of calls to another number, holding pending calls for a given number in a "stack", automatic redialing of busy numbers and abbreviated dialing codes are a few that are offered besides the conventional conference calls, paging, control of incoming and outgoing calls and long distance dialing for each extension.

Figure 6-16. *PABX System.*
(Courtesy of Southwestern Bell Telephone Co.)

All electronic central office exchanges bring these advantages to the total telephone system. With such capability, the selection of the transmission options in *Figure 6-3* for a given communication is much easier. In many cases it is searched through by the system itself. More conversations can be carried on the same transmission trunk lines because electronic switching provides the switching speed required to make maximum use of the most advanced methods of time division or frequency division multiplexing. Significant cost savings can occur in equipment and materials if a transmission link can carry an increased number of calls.

ESS System Components

The basic components of an electronic switching system are shown in *Figure 6-17*. The central control function is performed by a digital computer which keeps track of which lines in which paths are busy and which are available. It also determines which lines are initiating calls and what is the optimum transmission path for the requested connection. The sequence of events that occur when a telephone is lifted off-hook to initiate a call will illustrate the function of the ESS components.

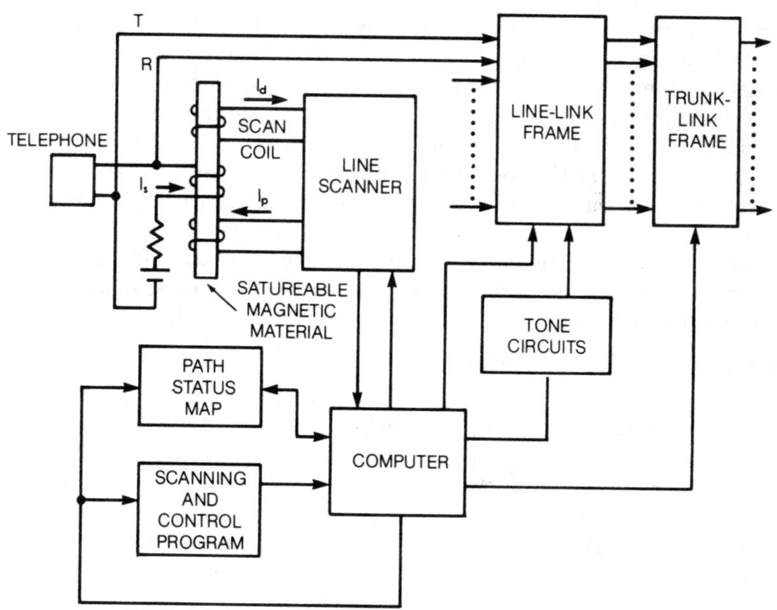

Figure 6-17. *Components of Electronic Switching Systems.*

The computer is continually scanning the lines coming into the ESS office to detect a new off-hook condition. It does this by pulsing a current I_p through a coil on a magnetic core containing another coil in series with the line to be scanned. If the telephone is off-hook and current is flowing through the line coil winding on this core, the core will be saturated and no current will be induced in the scan winding. In this case I_d will be zero and the system will know that the telephone connected to that core is off-hook (busy). If the phone is on-hook, I_s is zero and the scan pulse will induce a current pulse I_d in the scan winding indicating an idle phone.

The computer scans all cores in an orderly fashion and maintains a map of which lines are idle and which are busy. It also detects when a line changes from idle to busy, indicating the initiation of a call and sends a dial tone to the calling phone. It receives the dialed number from calling telephones and stores the number. Once the number is stored, it refers to a connection map and a path status map to determine what path should be used to connect the calling and called stations. Once this path is established, the called telephone is rung and a ringing tone is sent to the calling party. If the called party had been busy, the calling party would have been sent a busy tone. All of these tones are controlled and applied by the computer through its switching and tone circuits. When the called party goes off-hook, the computer maintains the talking connection until one of the parties hangs up. It updates its line status map constantly as these events occur.

The actual connection between the called and calling party is provided by electronic line-link and trunk-link matrix connections, much like those used in the Crossbar offices. Since a computer is involved in the control of an ESS office, a wide variety of the communications options discussed previously can be made available to all communicators in the system, and since the computer is programmable, these options can be changed easily depending on the system required. All of these features make the ESS office highly versatile and a trend for the future.

Frequency Division Multiplexing

As mentioned previously, as ESS system with its high-speed switching allows more conversations to be placed on a particular transmission link. One technique that was discussed in general in Chapter 4 that allows many conversations to share the same wire-pair or cable is that of frequency division multiplexing, FDM. A sample use of this technique placing 12 voice channels on a single wire or cable is shown in *Figure 6-18.* Each voice channel requires a maximum 4-kHz bandwidth as was shown in *Figure 6-10.* Each of the 12 voice channels will modulate a different carrier frequency spaced 4 kHz apart. The resulting separate bandwidths are summed so the 12 channels can be stacked on top of each other in the frequency spectrum. In *Figure 6-18*, the 12 carrier frequencies being modulated are 8140 kHz through 8184 kHz, causing the 12 voice channels to occupy non-overlapping frequencies from 8140 kHz to 8188 kHz.

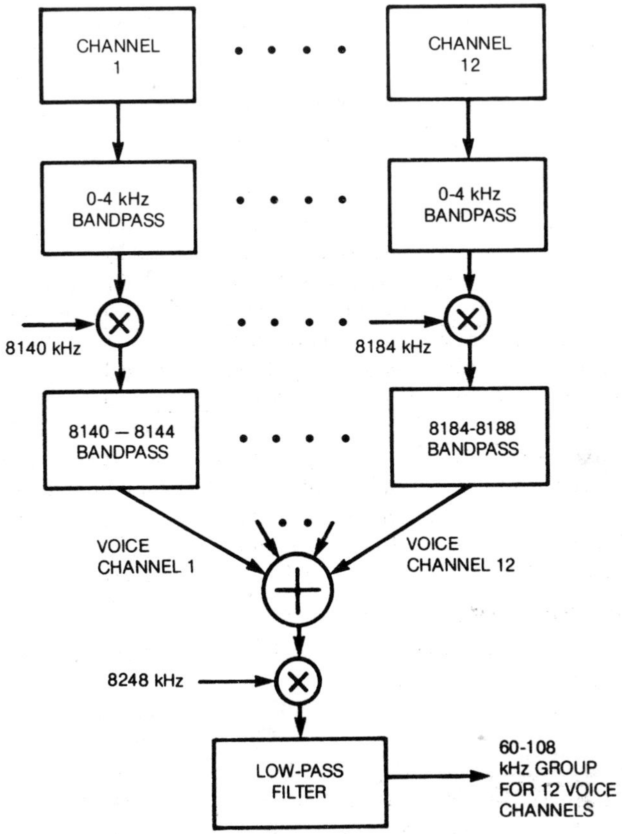

CHANNEL 1
CHANNEL 12
0-4 kHz BANDPASS
0-4 kHz BANDPASS
8140 kHz
8184 kHz
8140 – 8144 BANDPASS
8184-8188 BANDPASS
VOICE CHANNEL 1
VOICE CHANNEL 12
8248 kHz
LOW-PASS FILTER
60-108 kHz GROUP FOR 12 VOICE CHANNELS

Figure 6-18. *Basic Frequency Multiplexing of Voice Channels.*

The effect of this operation is shown in *Figure 6-19*. By further multiplying this spectrum with an 8248-kHz sinusoid and passing the result through a low-pass filter, this range of frequencies can be moved down to the 60-kHz to 108-kHz range, shown in the third plot of *Figure 6-19*. This bandwidth containing 12 channels is called a group. Such a 12-channel group could be sent along a single cable, allowing 12 conversations to share the same wire without interfering with one another. However, there is no need to stop here. Most wires can pass a much higher frequency range than 108 kHz; therefore, the process can continue so that even more voice channels can be sent over one transmission link.

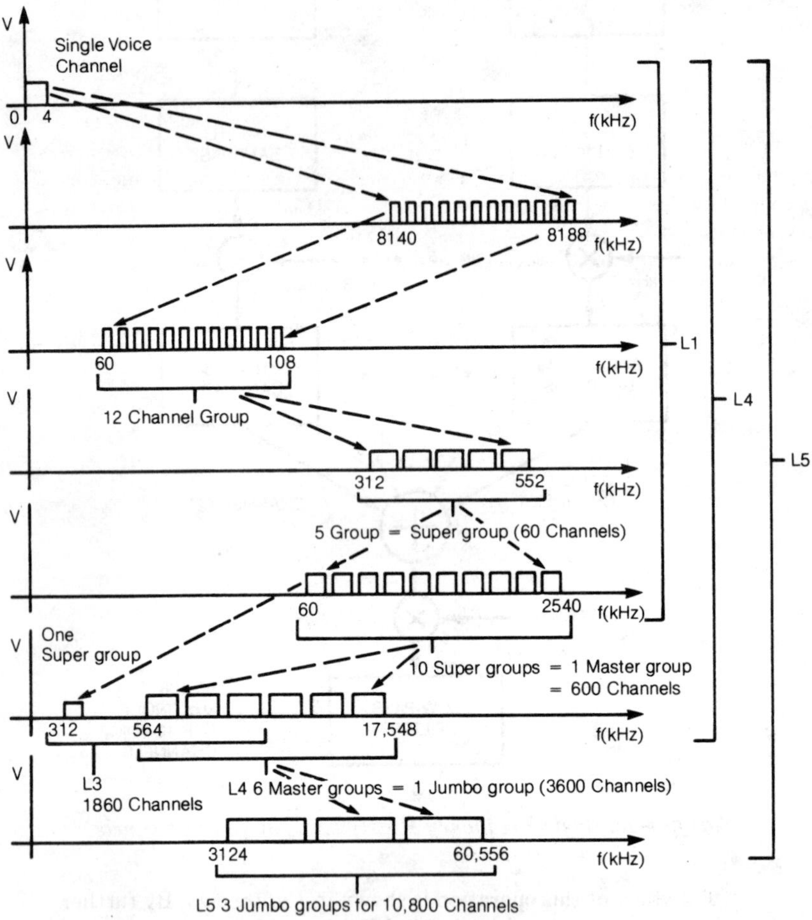

V
Single Voice Channel
0 4
f(kHz)
8140
8188
60
108
12 Channel Group
312
552
5 Group = Super group (60 Channels)
60
2540
One Super group
10 Super groups = 1 Master group = 600 Channels
312
564
17,548
L3 1860 Channels
L4 6 Master groups = 1 Jumbo group (3600 Channels)
3124
60,556
L5 3 Jumbo groups for 10,800 Channels
L1
L4
L5

(Summary of information presented in "Basic Carrier Telephony", D. Talley, Hayden Book Co., 1977, pp. 183-188)

Figure 6-19. *Frequency Multiplexing Carrier Systems.*

The fourth plot of *Figure 6-19* shows the effect of modulating five different carrier frequencies separated by 48 kHz (from 312 kHz to 552 kHz) with five different groups. These five groups containing 60 voice channels are stacked in a frequency range of 240 kHz in what is called a super group. By modulating 10 different carrier frequencies with 10 different super groups, the ten super groups can be stacked in a frequency range from 60 kHz to 2540 kHz. Such a combination, shown in the fourth plot of *Figure 6-19* and containing 600 voice channels, is called a master group. Such a bandwidth is within the capability of twisted wire pairs with repeaters spaced every eight miles. This carrier system of transmitting a master group along a single wire pair is called the L1 system. L1 cables provide three such two-way paths for a total capability of 1800 voice channel conversations between central offices.

System Capabilities

The basic characteristics of the L1 frequency division multiplexing scheme are summarized in *Figure 6-20.* If the repeater spacing is halved to four miles, the wire pair can handle a bandwidth of three master groups plus one super group for a total of 1860 voice channels per wire pair. This system is called the L3 system. Each set of L3 cables contains five such pairs devoted to two-way communications for a total of 9300 two-way voice channels over the same L3 cable. By using coaxial cables and repeaters spaced at two miles, each coax can handle six master groups, which is called a jumbo group, for a total of 3600 voice channels per coax. The L4 system provides an overall cable with nine such coaxial cables devoted to two-way transmission for a total of 32,400 voice channels between offices.

Carrier System	Voice Channels*	Repeater Spacing (miles)	Approximate Bandwidth (kHz)
L1	1800	8	2500
L3	9300	4	8500
L4	32,400	2	17,500
L5	108,000	1	60,500
WT4	230,000	15	3,000,000

*Number of Channels per Cable times the Number of Cables

Figure 6-20. *Telephone System Carrier Capacities.*

In the L5 system, three jumbo groups are assembled around carrier frequencies in a frequency range from 3124 kHz to 60,556 kHz for 10,800 voice channels per coax. The L5 line provides 10 two-way coax pairs with this frequency bandwidth capacity using a repeater spacing of one mile for a total of 108,000 two-way voice channels per L5 Line. Of course, each central office must have the electronics to provide the modulation equipment to assemble the groups, super groups, master groups, and jumbo groups and to disassemble these bandwidths at the receiving office. This increases the cost of the central office equipment. Offsetting this is the trememdous increase in inter-office capacity and the tremendous saving in wire and line costs between offices. Also, all of these frequency division multiplexing (FDM) techniques can be used with microwave links and satellite links. As an example, *Figure 6-20* lists the capacity of the WT4 helical waveguide system with a repeater spacing of about 15 miles as 230,000 simultaneous two-way voice channel communications.

Time Division Multiplexing

While frequency division multiplexing (FDM) has been used to great advantage in increasing system capacity without unacceptable increases in system costs, the use of time division multiplexing (TDM) offers even greater promise of system improvements. In order to use time division multiplexing, the transmissions between offices must be digital in nature instead of the analog information that is characteristic of telephone conversations. This means that the speech generated at the telephone must be converted to a digital signal, transmitted, and then reconverted into analog form to generate the speech at the called telephone.

The system components required are shown in *Figure 6-21*. The components labelled A/D are analog-to-digital converters of the type discussed in Chapters 2 and 3 where input speech is converted to an 8-bit binary code through pulse code modulation. These conversion circuits are located at the central office with the standard circuits used to locate the calling phone and to provide a dial tone. In the central office, the 8-bit code is sent in serial form to a component called a multiplexer. This circuit, as the result of an input control code, selects one of many inputs and places that input on the one output line. In *Figure 6-21*, the multiplexer shown is an 8-line-to-1-line multiplexer. Normally in a central office this will be an n-line-to-1-line multiplexer with n governed by the central office capability. The output from the multiplexer is the trunk line out of the central office to the remote central office of the called telephone.

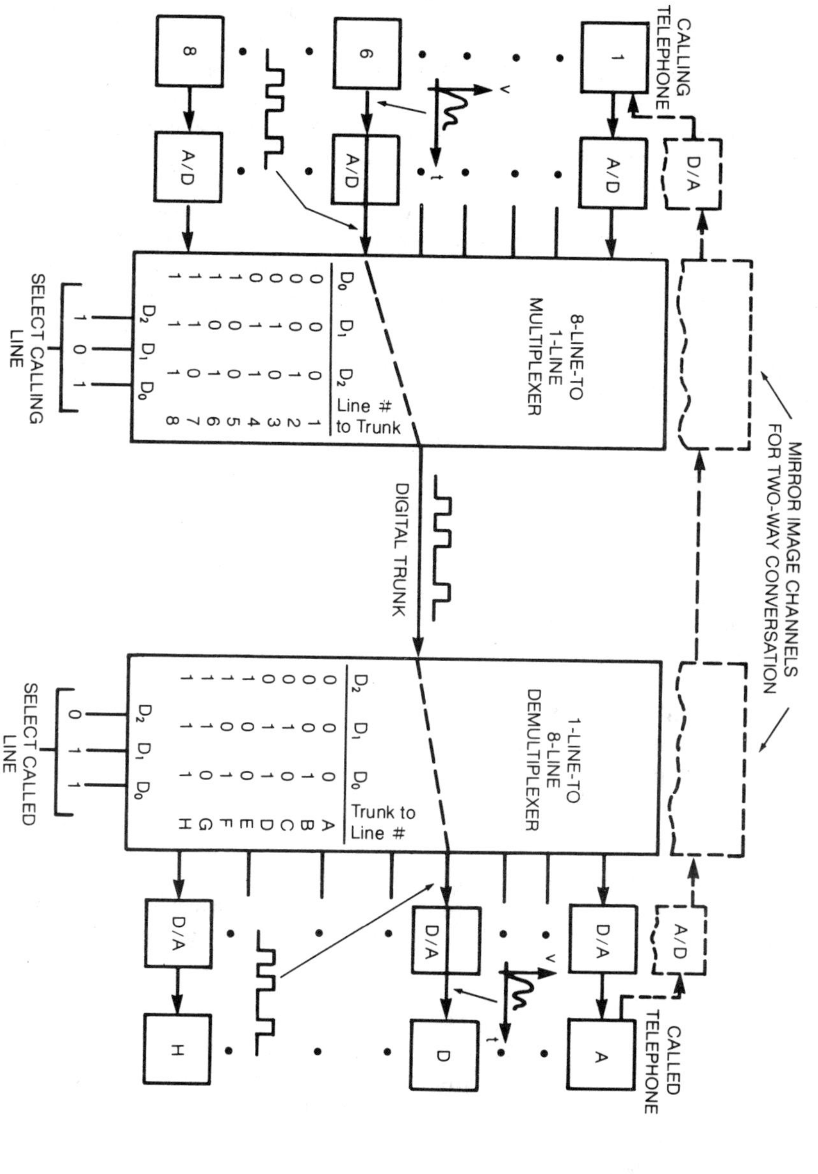

CALLING TELEPHONE

1 A/D D/A

6 A/D

8 A/D

8-LINE-TO 1-LINE MULTIPLEXER

D_0	D_1	D_2	Line # to Trunk
0	0	0	1
0	0	1	2
0	1	0	3
0	1	1	4
1	0	0	5
1	0	1	6
1	1	0	7
1	1	1	8

D_2 D_1 D_0

1 0 1

SELECT CALLING LINE

DIGITAL TRUNK

MIRROR IMAGE CHANNELS FOR TWO-WAY CONVERSATION

1-LINE-TO 8-LINE DEMULTIPLEXER

D_2	D_1	D_0	Trunk to Line #
0	0	0	A
0	0	1	B
0	1	0	C
0	1	1	D
1	0	0	E
1	0	1	F
1	1	0	G
1	1	1	H

D_2 D_1 D_0

0 1 1

SELECT CALLED LINE

D/A A A/D

D/A D

D/A H

CALLED TELEPHONE

v t

Figure 6-21. *Basic Electronic Switching of Digital Signals.*

The code used to select the input line that is multiplexed onto the trunk line is determined and made active by the calling telephone. In the specific example of *Figure 6-21* the calling telephone is number 6. When the central office detects number 6 as the calling telephone, it provides a control code of 101 to the multiplexer to select the input line from calling telephone 6 so that it is placed on the trunk line to the remote central office of the called telephone. As the caller speaks, the equivalent digital code goes out serially onto the trunk line.

At the remote central office there is a similar arrangement of components to convert the received signal back to the analog speech signal. It is assumed that the called telephone number has been identified and that the appropriate switching and testing has occurred to determine that the called telephone is not busy. The trunk line serves as an input to a component called a demultiplexer. Again this would be a 1-line-to-n-line demultiplexer based on the capabilities of the remote central office. For *Figure 6-21* it is a 1-line-to-8-line demultiplexer.

The code to select the output line (011 in *Figure 6-21*) has been set by the called telephone number and thus, the trunk line is connected to the called telephone (D in *Figure 6-21*). In the path from the demultiplexer to the called telephone is a D/A converter which accepts the serial code and generates the original speech signal by the calling party to excite the speaker of the called telephone (see Chapter 3). Thus, conversation is possible from calling party to called party. As shown in *Figure 6-21*, a similar mirror image path would exist from the called party back to the calling party, so that normal two-way conversation can occur. Of course, eight lines is a trivial example in comparison with the 10,000 lines connected to each central office, but it does illustrate the basic switching that occurs. The type of switching illustrated in *Figure 6-21* is called space multiplexing, since it determines the physical path used for transmitting the signal.

T1 TDM System

Once the switching system is capable of handling digital signals, it becomes possible to use time division multiplexing (TDM) to send several conversations on the same wire pair or cable. The basic approach used in the simplest system, the T1 TDM scheme, is shown in *Figure 6-22*. Each voice channel is assigned a time slot. In *Figure 6-22*, channel 1 is assigned the first 8-bit time segment, channel 2 is assigned the next 8-bit time segment, and so on with channel 24 assigned the last 8-bit time segment. A final bit, the framing bit, is used to synchronize the system. Since an analog signal needs to be sampled at twice its highest frequency to obtain an accurate digital representation of the information content of the signal, and since each channel has a maximum bandwidth of 4 kHz, the voice channel must be sampled at 8,000 hertz. This means that each 8-bit sample must be re-examined every 125 microseconds.

For 24 channels, 193 bits (24 × 8 + 1 framing bit) occur 8,000 times a second for a bit rate of 1,544,000 bits each second. This is written as 1.544 megabits/second. This bit rate can be handled by channels connected by wire pairs with repeaters spaced at about one-mile intervals. The 1.544 megabits/second compares with a limit of around 2400 bits/second that can be sent on a standard analog voice band channel using FSK techniques. Thus, if digital computer data or television or other non-voice information must be transmitted, doing it on digital channels offers much more versatility and much higher speed transfer capability than an analog channel does. Further, the digital signal is much more immune to channel noise than is the analog signal.

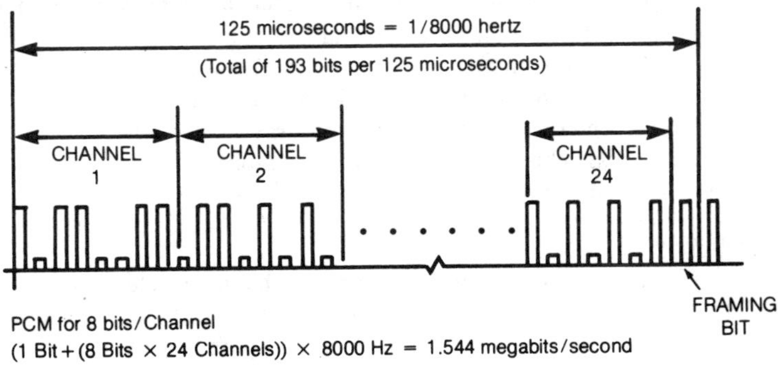

Figure 6-22. *Basic Time Division Multiplexing for T1 System.*

Distribution

In the T1 system, a three-level distribution system is used to accomplish the TDM. With this technique, 24 voice channels are sent on each T1 wire pair. The system structure is only slightly more complex than that used for systems that are not time division multiplexed. *Figure 6-23* indicates the additional components needed to implement TDM switching. It is assumed that space switching has occured so that all the input telephones of *Figure 6-23* are active calling telephones and all the output telephones are active called telephones. Time division multiplexing occurs by connecting the bits from each telephone that is calling onto the trunk in a sequence of the type shown in *Figure 6-22*. First the calling speech information must be converted to 8-bit digital form, once each 125 microseconds. This is done by the A/D converter. Each of the bits from the 8-bit codes produced by the A/D converters for each channel are outputted in series onto an input line to the channel multiplexer. This is accomplished by a modulo eight counter (counts to eight and repeats) that steps through each of eight different codes to select the particular bit that is placed on the multiplexer input line at a 1.544-megahertz rate. When each eight code count is finished, the modulo eight counter sends a pulse to the modulo 24 channel counter (counts to 24 and repeats) and advances it one step.

The code from the channel counter is selecting the particular channel input of 8 bits that appears on the trunk line starting with channel 1. Therefore, with the channel counter code of 00000, channel 1's 8-bit code will be placed on the trunk line in sequence in position 1 of the signal sequence shown in *Figure 6-22.*

The next time the channel select counter steps to code 00001, channel 2's input to the multiplexer will be placed in postion 2 of *Figure 6-22.* In like fashion, all 24 channels will be placed in sequence on the trunk line for transmission to the remote central office. Every 125 microseconds the pattern changes to reflect the equivalent codes for the information that is to be transmitted from calling party to called party.

Previous to the conversation information arriving at the remote central office, the dial pulse information has been distributed so that the calling and called telephones can be located and the correct active called telephones connected in the correct positions on the outputs to the demultiplexer at the receiving end. This requires a great deal of coordination and synchronization which is accomplished by the distribution control computer shown in *Figure 6-23.*

Demodulation

The reverse conversion must occur at the called telephone's central office. The digital trunk signal, synchronized to the calling office, places the positioned 8-bit codes onto the demultiplexer outputs in sequence as determined by the channel select code. The serial pulses are timed by a similar modulo eight counter as for the transmitting office. After the 8-bit code is latched in temporary storage, is is read out to the D/A converter every 125 microseconds and sent to the called telephone. Each output is sequenced through by the channel select code. Thus, the 24 input channels are sequenced out to the 24 output channels. The respective correct called telephone has been connected previously to the correct channel by the distribution control computer. If calling telephone 1 has been placed in channel 1 position as its input, then telephone A is the called telephone dialed by the calling party at telephone 1 (*Figure 6-23*).

In reality, different types of data may be coming over the channels. Some of the channels may be carrying voice conversations while others may be carrying computer data. Further, these time allotments can be assigned on a demand or as-needed basis; that is, they could be dynamically allocated. Finally, the techniques of TDM, FDM, and space switching can be combined to obtain the most capacity possible out of a given line, cable, or microwave/ satellite bandwidth.

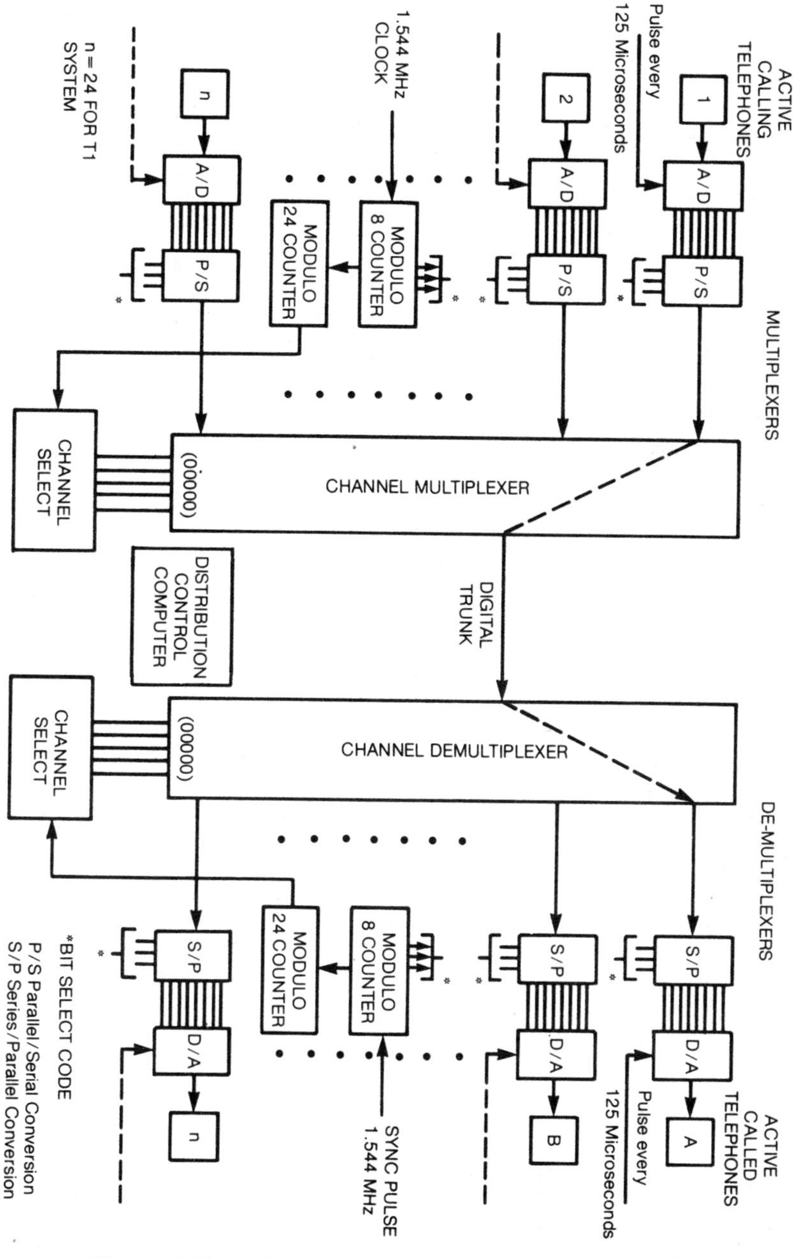

Figure 6-23. Basic Time-Division-Multiplexing Hardware.

SYSTEM DATA RATES

Figure 6-24 summarizes some of the data rate capability in modern telephone systems, including the data rates possible over standard voice channels using frequency-keying (FSK). The T2 system offers four times the bit-rate and voice-channel capacity of the T1 system. The T4 system offers data rates of 274 M bits per second per coaxial cable for 4032 voice channels per cable. With an 8-cable T4 line, over 30,000 voice channels can be transmitted. As will be seen in the next chapter, each T4 coax could alternatively be used to transmit three television channels.

System	Bit Rate (Megabit/ Second)	Voice Channels	Transmission Medium
Voice Band FSK	0.0024	1	Voice Grade Wire Pair
Wideband FSK	0.064	1	Voice Grade Wire Pair
T1	1.544	24	Twisted Wire Pair
T2	6.312	96	Twisted Wire Pair
T4	274	4032	Coax or Microwave Links
WT4	16,000	240,192	Helical Waveguide
Optical Fiber	900-54,000	12-80,000	Glass Fiber

Figure 6-24. *Digital Capacities of Telephone Systems.*

With the capacities of the L5, T4, and WT4 systems and the potential capacities of optical fiber waveguides, telephone systems will offer an ever expanding amount of communication capacity over a worldwide secure switching network for use not only with telephone conversations, but also with the entire range of society's communications needs. Many of these potential applications will be covered in the remainder of this book.

WHAT HAVE WE LEARNED?

- Telephone and telegraph communication systems provide a means of transferring information from one point to another for establishing individual-to-individual communication.
- Such networks are essentially huge switching systems connecting one communicator with another.
- Much of the signaling and control in telephone networks has been done with dc currents and electromechanical switches using sequential switching in the SxS system or matrix switching in the Crossbar system.
- Dial tones and electronic switching systems controlled by computers are the switching system of the future, offering low cost, high reliability, and expanding capabilities over electromechanical systems.
- Frequency division multiplexing and time division multiplexing offer techniques for sending many simultaneous conversations over the same physical path.
- In the future, digital signal processing and multiplexing will increase.

WHAT'S NEXT?

One of the most powerful communications techniques is that of television communication. Such communication can be broadcast through the air to home receivers. It can be sent over telephone systems to distribute programs and information nationally and internationally. It can be sent over private cable lines and via satellite to provide complete information, entertainment, and data services. This powerful and yet commonplace form of communication will be the subject of the next chapter.

Quiz for Chapter 6

1. Originally telegraph and telephone systems used the following signal transmission technique:
   a. radio waves.
   b. microwave links.
   c. cables for signal transmission.
   d. wires for signal transmission.
   e. c and d above.
2. If two telephones contain different first three digits of a 7 digit number:
   a. the call is to two different telephones at the same central office.
   b. the call is from one part of the country to another thousands of miles away.
   c. the call is to one of the 10,000 subscribers connected to the same central office that the calling party is connected to.
   d. the call is from one central office to a nearby central office.
3. Telephone switching offices provide the following switching functions:
   a. switching one local subscriber to another at the same central office.
   b. switching from one central office to another.
   c. switching from one central office to a toll center.
   d. All of the above.
4. A ringing current is an ac signal in the following frequency range (Hz)
   a. 0 to 20.
   b. 10 to 47.
   c. 20 to 47.
   d. 1000 to 1250.
5. The frequencies generated by pushing button 9 in a touch-tone telephone would be:
   a. 697 and 1209 hertz.
   b. 852 and 1336 hertz.
   c. 941 and 1477 hertz.
   d. 852 and 1477 hertz.
6. Speech bandpass amplifiers limit frequencies on a telephone channel to the following range in hertz.
   a. 0 to 4000 hertz.
   b. 0 to 3400 hertz.
   c. 100 to 4000 hertz.
   d. 200 to 3400 hertz.
7. A rotary dialing mechanism pulses the line at the following rate in pulses per second:
   a. 5
   b. 10
   c. 15
8. An electronic switching office keeps track of available communication paths and the status of all telephone lines and requested lines with:
   a. large memories.
   b. a central computer.
   c. large banks of stepping switches
   d. a and b above.
9. If a television channel requires 90 megabits/second data rate, how many television channels could be sent over two T4 coaxial cables?
   a. 3
   b. 1
   c. 6
   d. 10
10. If high fidelity stereo music requires a bandwidth of 16000 hertz per channel and 8 bits of PCM accuracy, how many channels of the T1 system would be required to transmit this information?
    a. 8
    b. 16
    c. 24

(Answers in back of the book)

Radio and Television Systems

ABOUT THIS CHAPTER

In the last chapter the basic concepts of the familiar telephone system were covered. An equally familiar system that is taken almost as much for granted as the telephone system is radio and television. Most people are never very far from a radio or television, and many hours per day are spent being entertained or informed by one or the other of these systems. Since television includes radio-type information in its sound portion, and since both depend on broadcasting information over an electromagnetic radiation transmission link, an analysis of the basic concepts of television will also serve to cover the fundamentals of radio. Thus, this chapter will emphasize the techniques and applications of television systems, and at the same time, various discussions will indicate how these techniques apply to radio broadcasting. The first most basic concept of understanding television systems is the technique for transmitting pictures electronically.

HOW ARE IMAGES TRANSMITTED ELECTRONICALLY?

The components required to broadcast pictures from a television camera to a television receiver and display them on a TV picture tube are shown in *Figure 7-1*. The purpose of the camera is to convert the optical information (reflected light waves) of the image into corresponding electrical signals. The microphone converts any sound associated with the image, such as speech or music, into corresponding signals. These signals must be modulated onto a high-frequency carrier and amplified until they are powerful enough to cause electromagnetic radiation from the broadcasting antenna. The range and direction of such radiation depends on the power of the transmitter, the frequency of the carrier, and the design of the antenna. These factors were discussed in some detail in the chapter on the communications spectrum. If the receiver is properly located within range of the transmitter, it will detect the radiation as a current flow in its antenna. At the receiver, this signal level is increased with a high-frequency amplifier, demodulated, and processed to produce the original sound in a speaker and the original image on a picture tube. It is this processing, both at the camera and at the television receiver, that must be understood if the operation of a television system is to be understood. As shown in *Figure 7-1*, the transmission link many times is by cable rather than electromagnetic radiation. This in no way changes the system concepts.

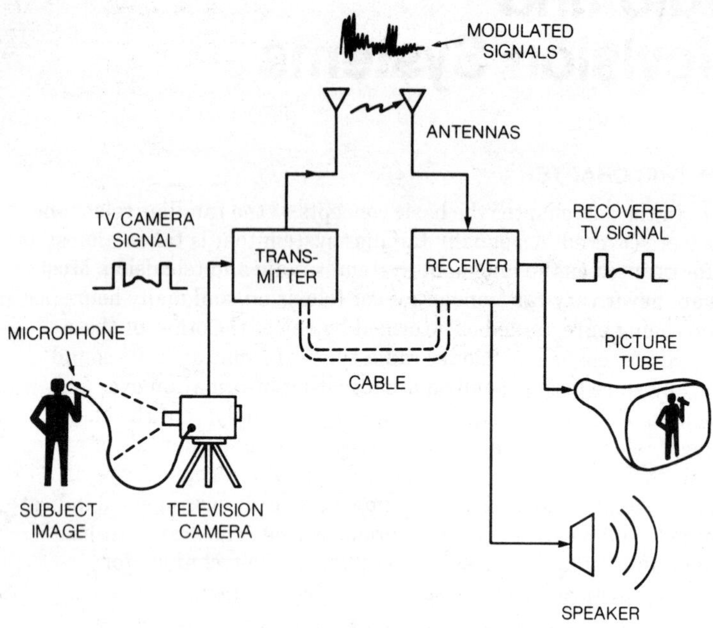

Figure 7-1. *Basic Elements of a Television System.*

Camera Image to Electrical Signals

Figure 7-2 illustrates how a television camera converts an optical image into a corresponding electrical signal. In this case, the image is that of the letter H. Lenses in the front of the camera focus this image on the face of the camera tube, which is about an inch in diameter. Enclosed in a vacuum inside this tube is a photosensitive material covering the inside of its face, a sharply focused beam of electrons that scans across a target in the pattern shown, and some means for detecting current variations in the electron beam as it scans across the target. When light hits the face of the tube and the photosensitive material, electrons are emitted, but when the face is dark no electrons are emitted. The emitted electrons cause a copy of the image to appear on the target surface. However, now the image is defined by the amount of positive charge instead of negative charge. A high positive charge for high intensity light; a low positive charge for low intensity or nearly dark conditions.

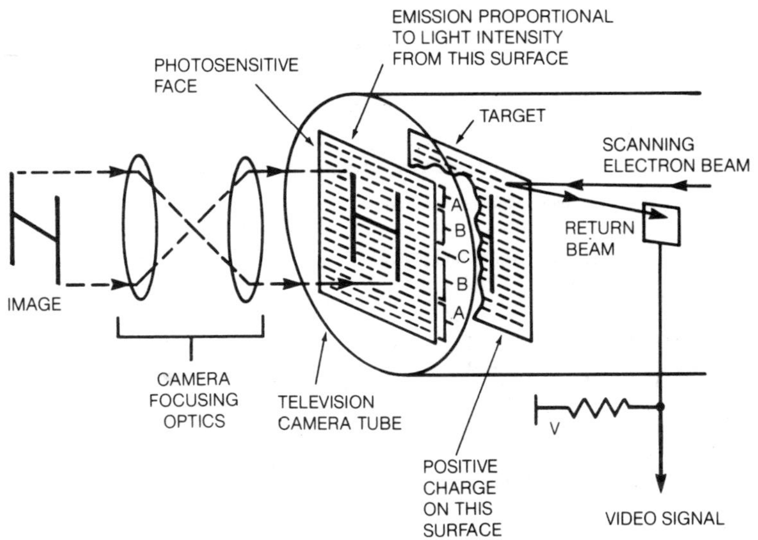

Figure 7-2. *Conversion of Optical Image to Electrical Signals.*

The Scan Pattern

As the electron beam is caused to scan the target from left-to-right across the tube for one horizontal line, it is moved slowly down vertically in a continuous manner causing it also to scan toward the bottom of the tube. As the beam scans, a reflected beam is collected and detected through a resistor shown in *Figure 7-2.* The reflected beam will contain the same number of electrons as the original scanning beam if there is no positive charge on the target surface. If there is a positive charge the reflected beam will have less electrons (less current) because some electrons combine to neutralize the positive charge. Therefore the current in the reflected beam will be inversely proportional to the light intensity of the image that is focused on the camera tube. The current variations produce the voltage variations of *Figure 7-3* at the output of the television camera tube.

As shown in *Figure 7-3a,* the voltage is greater (current is a maximum) when there is little light focused on the face of the tube, and it is a minimum when there is white light focused on the face of the tube. When there is total darkness the voltage will be at a maximum.

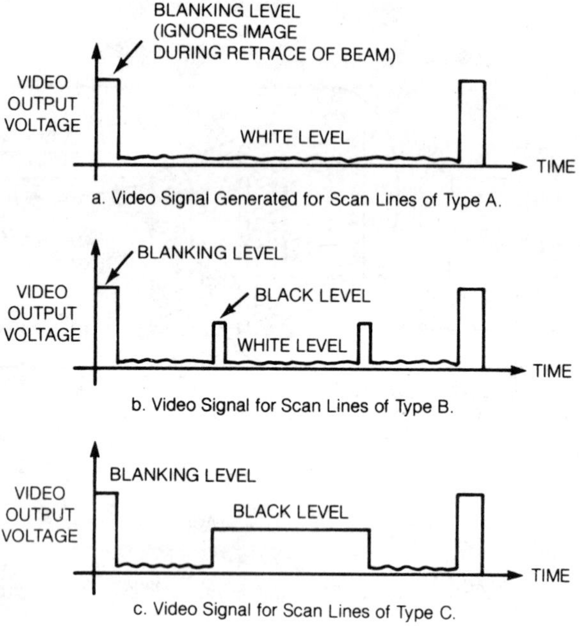

Figure 7-3. *Conversion of Optical Image to Electrical Voltage Variations.*

Scanning a Character

When the scanning electron beam completes a scan line, it returns to the left side of the face of the tube and starts another scan line. This return is called flyback. The beam current is ignored during flyback because the output voltage is set to its maximum (total darkness) during this flyback time *(Figure 7-3)*. As the beam continues to scan, the three signals that are generated as horizontal line information at various points on the letter H are shown in *Figure 7-3.* When no part of the letter is covered by the scan, the signal stays at its white level between flyback pulse levels. This is shown in *Figure 7-3a.* When the scan line begins to encounter the H, then two dark pulses occur, as shown in *Figure 7-3b*, when the beam crosses the dark legs of the H image. Otherwise the signal stays at the white level. During flyback the voltage is made even more positive than the level that corresponds to black in the image. Finally, when the horizontal scan line covers the horizontal mark of the letter H, the output voltage will be at the black level for the entire width of the letter as shown in *Figure 7-3c.*

The scan continues until it reaches the bottom of the face of the tube. As a result, the entire two-dimensional image is converted into a long continuous electrical signal with time containing as many flyback pulses as there are horizontal lines scanned in the two-dimensional field. Between two of the flyback pulses, the white-to-black variations in the image will be converted to corresponding voltage variations (low to high) at the output of the tube. The flyback pulses tell the receiver picture tube electronics when to cause the beam to flyback to the left side and when to start a new line. The electrical signal (called the video signal) between the flyback pulses tells the receiver what variations of white and black to produce on a given horizontal line of the regenerated picture. The electronic circuits in the camera tube (transmitter) and the picture tube (receiver) must cause the electron beam to scan in the desired pattern and must be synchronized together.

SCANNING THE RECEIVER PICTURE TUBE

The pattern that is scanned is illustrated on the television receiver picture tube in *Figure 7-4*. The video signal modulates the intensity of the electron beam. When the light intensity is bright at the camera, the beam will cause a bright spot on the picture tube at the same location. A changing current in the deflection coils produces a magnetic field which causes the beam to scan the line traces that correspond to the scan pattern at the TV camera. This is called the raster. The electron stream generated by the electron gun is focused into a beam to produce a fine spot on the picture tube face. The picture tube face contains a light emitting phosphor which emits light in proportion to the intensity of the electron beam current. The beam intensity is proportional to the video signal voltage applied between the control grid and the cathode of the tube.

Reproducing a Character

The signal patterns of *Figure 7-3a, b,* and *c* when applied to the picture tube reproduce the letter H on the picture tube. When the beam flies back from right to left, the flyback pulses cut-off all beam current, causing no light from the picture tube face during this retrace. For those lines not covering any part of the letter H *(Figure 7-3a)*, the low voltage for white allows the maximum beam current along the entire line, which provides a white horizontal line. For those traces of the form of *Figure 7-3b*, the line will be white except for two narrow portions which are black. For the traces that result from *Figure 7-3c*, most of the line will be black except for the area outside of the H. Thus, the video signal applied to the picture tube along with the appropriate deflection voltages will reproduce the image converted to electrical form by the television camera.

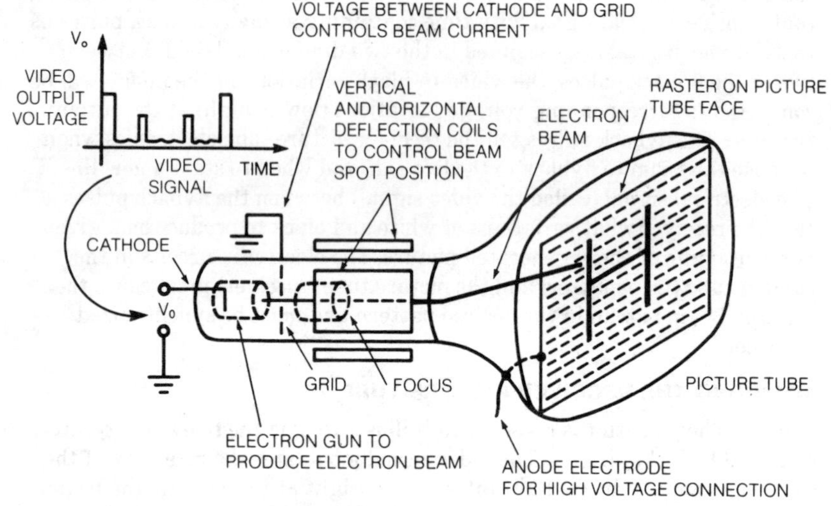

Figure 7-4. *Reproduction of Television Image on Picture Tube.*

Gray Scale and Deflection Waveforms

The video signals discussed thus far have been three level – black, white, and beam cut-off. The video signal for a picture with a wide range of gray tones would have many voltage levels as shown in *Figure 7-5a*. The horizontal and vertical deflection voltages required to cause the beam to scan the required pattern are indicated in *Figure 7-5b* and *5c*. There are several important details indicated by these signals. First of all, as mentioned previously, there are now many levels of gray in the picture video signal between black and white levels. As the picture information changes, these levels vary continuously. Second, small pulses are shown on top of the flyback (blanking) pulses. These are synchronization pulses to make sure the receiver horizontal and vertical deflection signals *(Figure 7-5b* and *5c)* are synchronized with those at the transmitter.

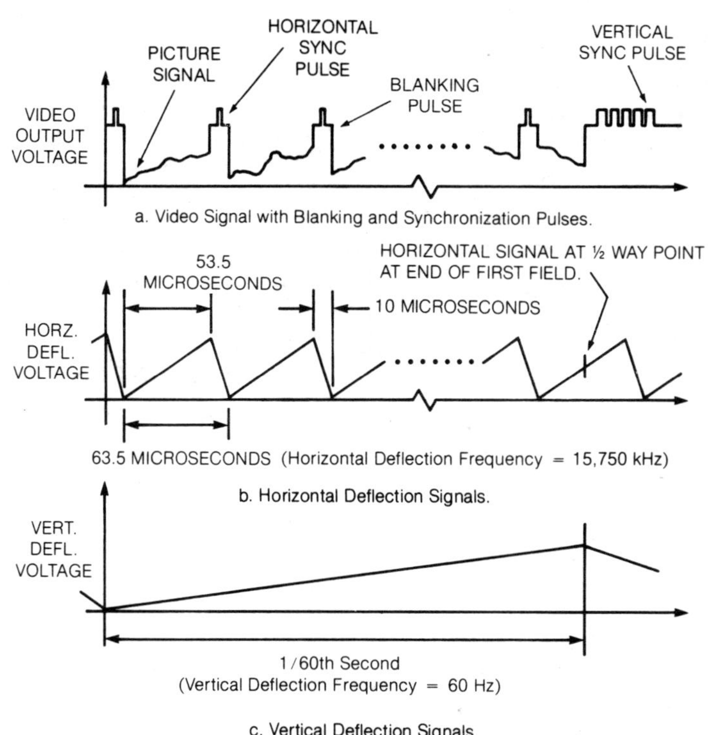

Figure 7-5. *Video and Deflection Signals in Television Systems.*

Horizontal and Vertical Scan Frequencies

The frequency of the horizontal deflection signal and the flyback pulses is 15,750 cycles per second to provide a spacing between corresponding portions of the flyback or horizontal deflection signals of 63.5 microseconds. Of this 63.5 microseconds, approximately 10 microseconds is the width of the flyback pulse and thus 53.5 microseconds is the available video signal time per horizontal line. The horizontal trace at the end of the vertical deflection (bottom of the picture) occurs half-way across the picture. The period of the vertical deflection signal is 60 hertz. These periods and frequencies cause the actual raster scan pattern to look like that of *Figure 7-6.*

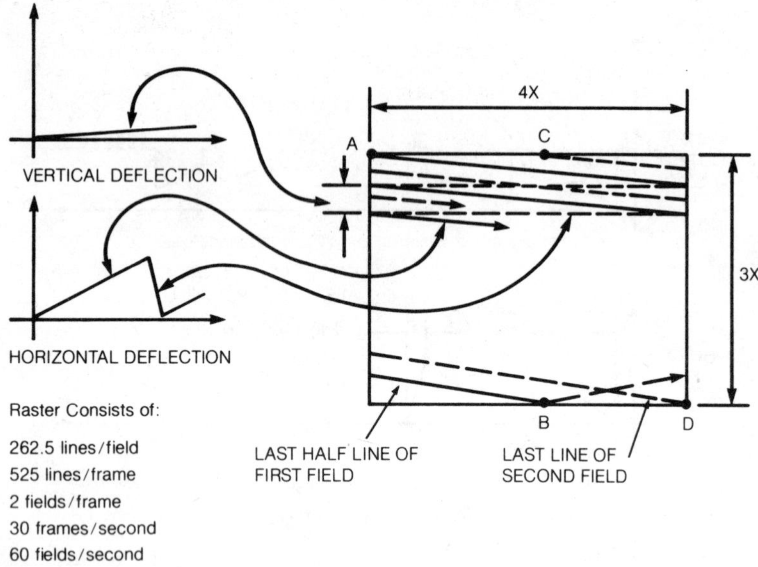

Figure 7-6. *Interlaced Scan Pattern for Television Pictures.*

Interlaced Scan Pattern

The solid lines show the scan pattern starting at A for the first field of the picture, which ends at the bottom of the picture area (B) one-half way across the picture width (recall the vertical and horizontal signal relationships of *Figure 7-5)*. Thus, there are 262 1/2 lines per vertical trace or field. As the vertical retrace occurs, the horizontal deflection signal continues an even number of deflection periods so that the first line of the second field of the picture starts at point C. This field scan pattern continues, offset from the first field scan pattern so that an interlaced scan pattern occurs from field to field. This gives the effect of seeing the same picture 60 times a second, which effectively avoids any flicker in the picture. This type of scanning is called interlaced scanning with two interlaced fields of 262 1/2 lines each per frame. There are 60 fields per second and 30 frames per second. In all fields, the horizontal scan moves across the entire width of the tube while the vertical deflection moves down slightly, causing a small slant in the horizontal scan as shown in *Figure 7-6*. The flyback time is so short with respect to the vertical deflection rate that the flyback trace is almost straight across the tube face. As indicated, the aspect ratio of the picture is standardized at a 4-to-3 width-to-height ratio.

The 525 lines per frame provides a picture resolution of about 300 to 350 lines in the vertical direction and a comparable 400 or more lines in the horizontal direction for video frequencies up to about 4 megahertz. If one visualizes the picture broken down into many square elements such that there are 300 rows of such elements and 430 columns of such elements, there would be about 140,000 fundamental picture elements (called pixels or pels) per frame. This corresponds with about 125,000 pels for 16mm movie film, which means that the television picture has about the same resolution or sharpness of detail as would a 16mm movie.

VIDEO BANDWIDTH

The effect of horizontal resolution in lines and the required video bandwidth can be seen by examining *Figure 7-7a*. Assume that the square pels are of width d. The most content the picture can have is when every other pel is black, which causes a video signal of the form of *Figure 7-7b*. If the pel is of dimension d with a 2d spacing between black bars and L/d pels per line, the square wave in the video signal will consist of L/2d pulses in 53.5 microseconds. The period of the pulses is 53.5 microseconds divided by L/2d and the frequency of the resultant square wave is L/2d x 53.5 in MHz. Substituting n for L/d gives $f = n/107$ or $n = 107f$. With a given bandwidth this equation allows the resolution to be determined. For example, with $f = 4$ MHz, 428 pels per horizontal line would be possible. With the 4:3 aspect ratio the horizontal resolution would be equivalent to a vertical resolution of 300 to 350 lines. Obviously, the higher the bandwidth of the television video signal, the more the resolution of the picture. The concept of pels is important not only from the standpoint of resolution but for determining digital transmission requirements as well. These concepts will be discussed later in this chapter.

All of the discussion thus far has assumed a black-and-white (monochrome) television system. In order to transmit and reproduce color pictures, several elements must be added to the system.

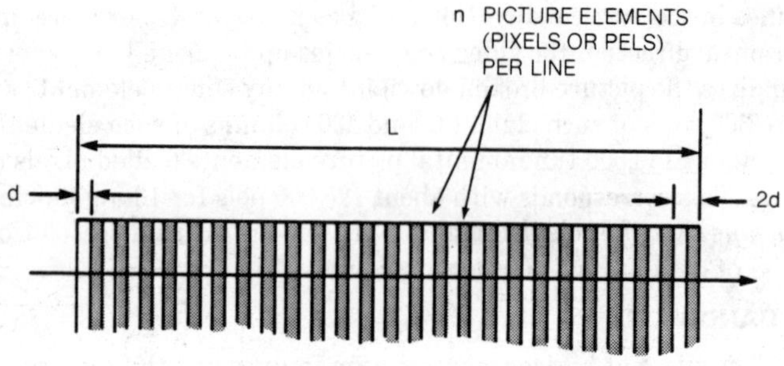

a. Black Bar Pattern with Minimum Spacing.

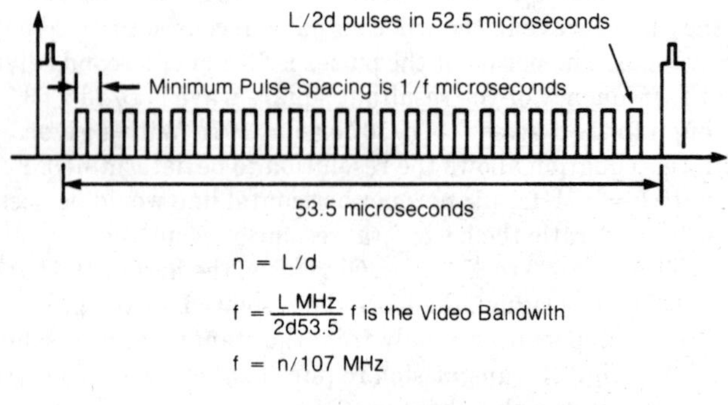

$$n = L/d$$

$$f = \frac{L \text{ MHz}}{2d53.5} \quad f \text{ is the Video Bandwith}$$

$$f = n/107 \text{ MHz}$$

b. Video Signal for Pattern of 7a.

Figure 7-7. *Pattern to Determine Resolution of Television.*

HOW ARE COLOR PICTURES PROVIDED ELECTRONICALLY?

In order to transmit the information contained in a color picture, the picture content must be broken down into its primary colors. In color television the primary colors chosen are red, green, and blue, since appropriate combinations of these components will produce any color. To identify the content of each of these primary colors in the picture, the television camera consists of three separate tubes or electron guns – one for each of the three colors (see *Figure 7-8*). The light from the scene viewed by the camera is passed through red, green, and blue filters and scanned by three separate electron beams.

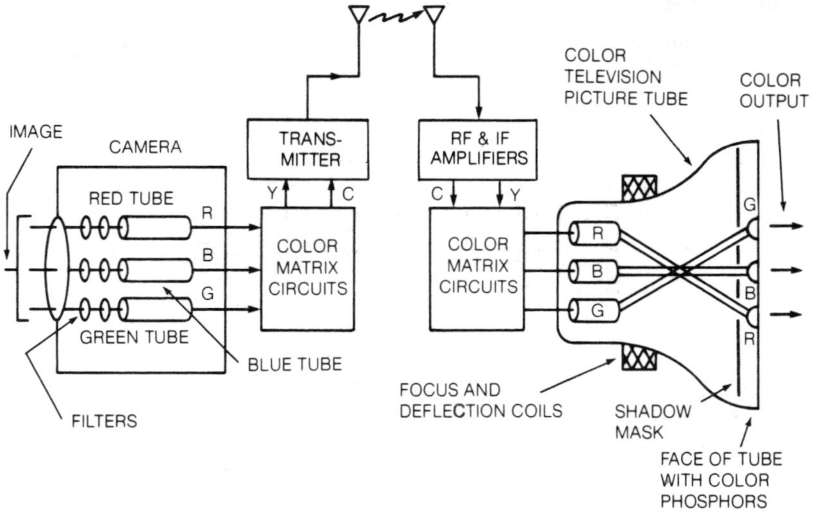

Figure 7-8. *Transmitting Color Images.*

Luminance and Chrominance

The signals for each color are combined by an electronic matrix circuit to produce two signals, called the luminance (Y) which contains the gray level information, and the chrominance (C) which contains the color information. When the luminance is applied to an ordinary black-and-white tube, a normal black-and-white picture results.

The transmitted signal contains the luminance and chrominance signals. When the signal is processed by a color receiver the red, green and blue signals are recovered, amplified, and then modulate the intensity of an electron beam in the color television picture tube. The face of the tube now has three phosphors located in a regular pattern next to each other over the whole surface of the tube – one for red, one for green, and one for blue. In the tube, as shown in *Figure 7-8*, each of the colors has its own electron beam which is focused on the tube face so that it hits the color phosphor that corresponds to its color. A shadow mask plate is placed between the beams and the color phosphors so that the red beam can hit only the red phosphor, the green beam only the green phosphor, and the blue beam only the blue phosphor. The chrominance amplitude and phase determine the beam currents for the three color beams. The amplitude contains the color intensity (called color saturation) and the phase contains the blend (hue or tint) of color. The tube deflection system moves all three beams in unison through the interlaced scan pattern of *Figure 7-6*. The focusing electrodes for each beam provide for precise focus and aiming of the individual beams of the three-beam group. The overall result is the reproduction of the original color picture detected by the color camera.

Representing the Color Spectrum

It may seem impossible for a single chrominance sinusoid to contain all the color information in a color picture. However, if the color spectrum is thought of as a spectrum of continuously varying hues or colors from deep blue to deep red, it is easy to see how an electrical parameter, such as voltage, current, resistance or phase can be assigned values to represent all the colors. In color television the parameter chosen is the phase of the chrominance sinusoid. Recall that it was stated previously that phase is a measure of the position in time of a waveform compared to a reference.

Phase of a Color Signal

The basic idea is illustrated in *Figure 7-9*. *Figure 7-9a* is a reference sine wave signal. It crosses the time axis at zero (point A) when the amplitude is zero and is starting to increase to a positive maximum. The waveform in *Figure 7-9d* is said to be 180° out of phase because even though its amplitude goes through zero at zero on the time axis (point B), it is starting to increase to its most negative maximum. It is a mirror image of the waveform in *Figure 7-9a*. Another way of looking at *Figure 7-9d* is that it is like a *Figure 7-9a* waveform that has moved in time position one half of a cycle down the time axis (Point C has moved to the Point B position). One full cycle is 360°; one-half of a cycle is 180°. The waveform of *Figure 7-9c* is said to be 90° out-of-phase from *Figure 7-9* because it is like a waveform that has moved in time position by a quarter of a cycle (Point D has moved to the Point E position.)

Color Sync Signal

Figure 7-9a is a reference 3.58-megahertz sine wave signal called the color sync that is sent as part of the flyback pulse in the video signal so that the receiver knows what the transmitter called zero degrees phase. It is used to synchronize a 3.58-megahertz color oscillator used as a reference in the color receiver. The chrominance signal transmitted will be compared to this reference to determine the phase of the chrominance signal which determines the blend of color (the hue) in the camera signal.

The chrominance sine waves of *Figure 7-9b*, *7-9c* and *7-9d* are signals received at different times representing the color of the camera signal. These are contained in the picture signal and are transmitted at the same time as the luminance signals. The signal of *Figure 7-9b* is in-phase (zero degrees phase) with the reference of *Figure 7-9a*. This corresponds to a color of yellow. *Figure 7-9c* is 90° out-of-phase from the reference; it corresponds to a color between red and magenta. *Figure 7-9d* is a blue because it is 180° out-of-phase and a 180° out-of-phase electrical signal corresponds to blue. Thus, by detecting the phase of the chrominance signal with respect to the color sync, the color can be determined. Information is also contained in the magnitude of the chrominance. Detecting the magnitude sets the intensity of the color.

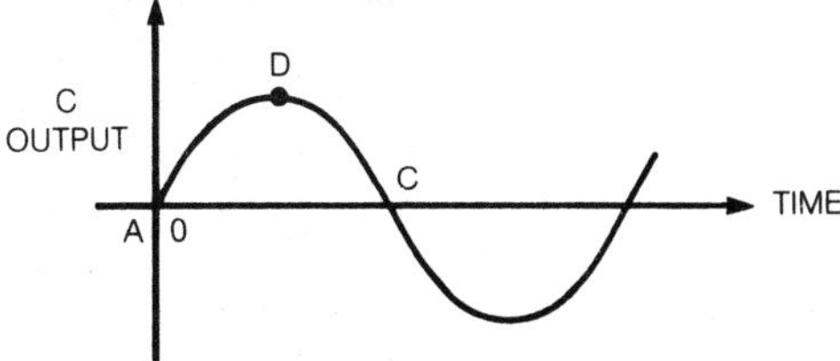

a. Color Sync Reference signal at 3.58 MHz.

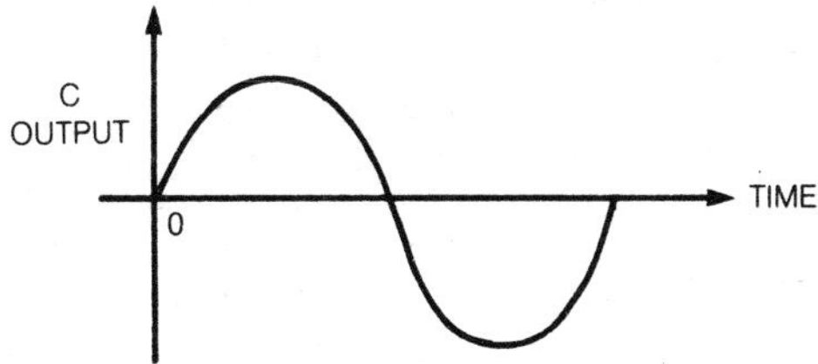

b. Color Signal (Chrominace) for Yellow.

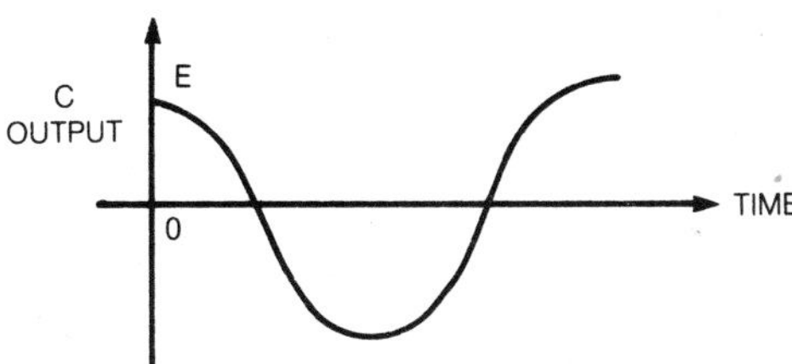

c. Color Signal for Red to Magenta.

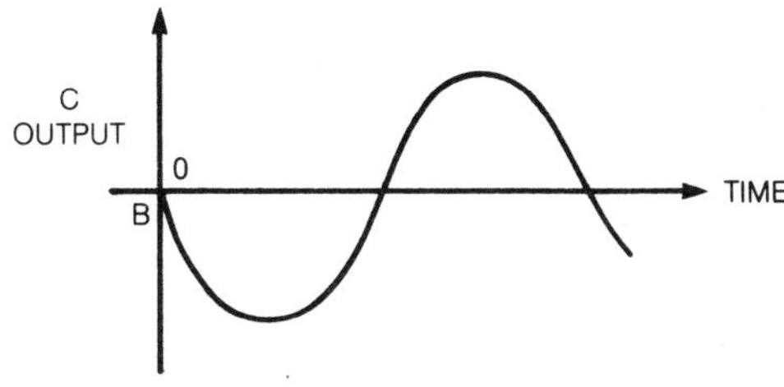

d. Color Signal for Blue.

Figure 7-9. *Relationship between Color Signal Phase and Color Reference.*

Separating Chrominance and Luminance

Different color television receivers use the chrominance and luminance signals in different ways. One of the more common approaches is to provide the color signals mixed with the luminance signals using the relationships expressed in the equations of *Figure 7-10.* The luminance signal is defined in *Figure 7-10a* as consisting of 30% of the red signal, 59% of the green signal, and 11% of the blue signal. When all three signals are of equal magnitude, this mixture will produce white, which is the proper result for the luminance signal. Now, instead of producing the three primary signals directly, the signals minus the luminance value are produced. By algebraically subtracting Y from one red magnitude the equation for the R-Y signal shown in a of *Figure 7-10* results. When all color signals are of equal magnitude, this R-Y signal is zero and only the Y signal exists. Similarly, by subtracting Y from B, the B-Y equation (b of *Figure 7-10*) can be determined. It also goes to zero when all three colors are present in equal amounts. The corresponding G-Y signal can be determined by subtracting Y from G or by combining the R-Y and B-Y signals as shown in c of *Figure 7-10.* These difference signals are useful for color television tubes since by applying the difference signal to one control electrode (a grid) and the Y signal to another (a cathode), the result is a beam current proportional to the color signal.

Luminance Signal:

$$Y = .3R + .59G + .11B$$

Color signals:

a. $R\text{-}Y = .7R - .59G - .11B$

b. $B\text{-}Y = .3R - .59G + .89B$

c. $G\text{-}Y = -.3R + .41G - .11B = -.51(R\text{-}Y) - .19(B\text{-}Y)$

Figure 7-10. *Basic Equations for Color and Luminance Signals.*

Color Circuitry

The way these difference and luminance signals are used in a color television can be understood by examining the color circuitry of a typical television. A typical color circuit block diagram is shown in *Figure 7-11*. The video signal previously received and amplified is fed into this group of circuits, including the blanking pulses for horizontal and vertical retrace. As shown in *Figure 7-11*, the signal also includes the color-sync burst on top of the flyback pulse, and the chrominance C sinusoid added to the luminance Y video signal. This signal is applied to the luminance low-pass amplifier, the chrominance bandpass amplifier, and the color burst bandpass amplifier. The two bandpass amplifiers pass the 3.58-MHz chrominance and burst signals but reject the luminance signal. The low-pass amplifier blocks the 3.58-MHz signals but passes the luminance signal. Thus Y is available out of the low-pass amplifier and is applied to the cathode of the electron guns as Y.

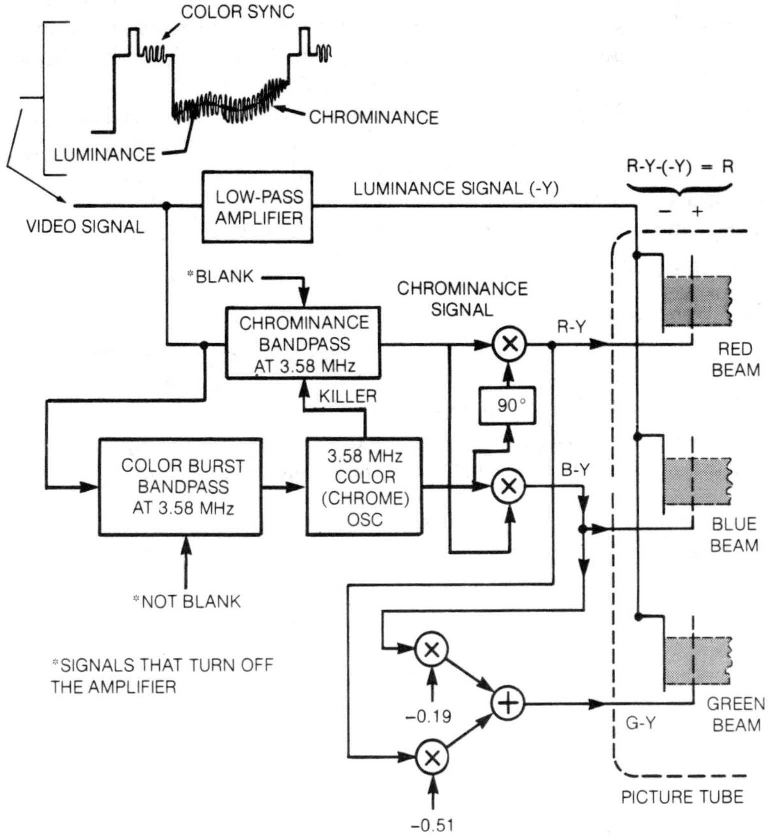

Figure 7-11. *Color Circuits in a Television Receiver.*

During blanking the burst amplifier is turned on and the chrominance amplifier is turned off. In this manner the color burst synchronizes the 3.58-MHz oscillator as the reference signal for the chrominance phase. During the horizontal scan the reverse is true. The burst amplifier is off and the chrominance amplifier is turned on to provide the chrominance signal output. The chrominance signal modulated with the color oscillator sinusoid delayed by 90 degrees produces the R-Y signals. Modulating the chrominance signal with an inverted color oscillator signal (180° out-of-phase) generates B-Y. The signal R-Y is multiplied by -0.51 and the signal B-Y by -0.19. These are added to produce G-Y. These three signals are applied to the grids of the three color guns. With the luminance signal (-Y) applied to the cathode, the net repelling voltage each electron beam sees is the grid voltage minus the cathode voltage. As shown in *Figure 7-11*, for the red gun this would be R-Y - (-Y) or R. The corresponding difference equations for blue and green vary the electron beam current in proportion to the appropriate color signals. As the beams excite the phosphors, the sum of the colors produced reproduces the original color at the appropriate spot on the receiver picture tube.

A color killer circuit is shown in *Figure 7-11*. Its purpose is to turn off the chrominance bandpass amplifier if no color sync is detected or if the color oscillator is defective. For this case, the only signal controlling the beams is Y (for all three beams if the receiver circuits are adjusted properly) and a standard black-and-white image is produced.

OVERALL RECEIVER

RF and IF Amplifiers

All television receivers, whether black and white (monochrome) or color must receive the signal from the transmitter via an antenna, amplify it and prepare it for display on the picture tube. Consequently the "front end" high-frequency (RF) and intermediate-frequency (IF) amplification stages including the video detector and the automatic gain control circuits (AGC) are common types to both receivers. An overall block diagram is shown in *Figure 7-12*. Television information can be broadcast on any of the channels shown in *Figure 7-13*.

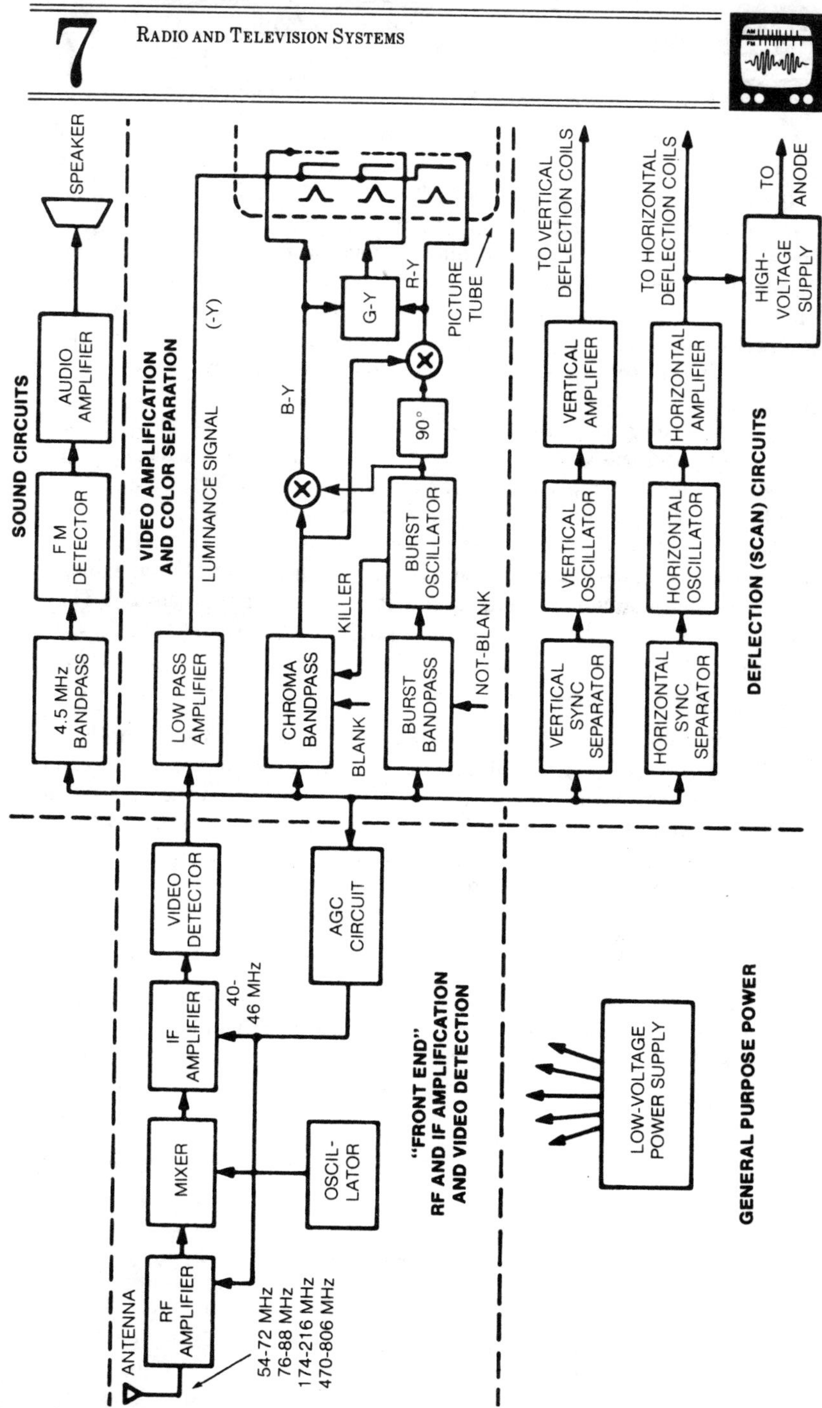

Figure 7-12. *Components of a Color Television Receiver.*

The RF amplifier is a bandpass amplifier that provides the initial amplification at the particular frequency received on the antenna. It passes only the frequencies of the channel selected by the receiver tuner or channel selector because the bandpass is set by this switch. The signal passing through the RF amplifier is mixed with a sine wave signal from an oscillator. The mixing produces a signal at the sum of the RF and oscillator frequency and another at the difference frequency between the two input signals. The channel selector tunes the oscillator at the same time as it tunes the RF amplifier so that the difference frequency out of the mixer is always 40 to 46 megahertz.

The selected channel signal will be amplified by the 40-to 46-megahertz bandpass amplifier called the intermediate-frequency amplifier or IF amplifier. IF amplifiers are used so that one high-performance amplifier section can be used to provide the majority of the receiver gain, regardless of the RF channel that has been selected. This allows the designers to produce a much better amplifier than would be possible if the amplifier had to provide a bandpass matched to the channel selected. The output of the IF amplifier has enough signal voltage to drive the video and sound circuits.

Video Detector and AGC

The video detector removes the 40-megahertz carrier from the amplitude modulated IF signal. The strength of the output of the detector is fed to automatic gain control circuits (AGC) to maintain the gains of the RF and IF amplifiers at desired levels over a fairly narrow range. This prevents strong antenna signals from behaving differently from weak signals at the output of the video detector. If a weak signal is received, the AGC control sets the RF and IF amplifier gains at a high value so that a strong video output is available. If the antenna signal is strong, the AGC control sets the amplifier gains low so that the video output is not overdriven or distorted. Thus, the AGC control circuit maintains a relatively strong undistorted video detector output that is largely independent of the strength of the antenna signal.

Sound Circuits

The video detector output is sent to four different receiver subsystems: the sound circuits, the video amplification and color separation circuits containing the luminance and chrominance amplifiers, and the deflection circuits. The luminance and chrominance circuits have already been discussed. *Figure 7-14* contains more detail of the frequency separation of the various signals in the video signal out of the detector. The sound subsystem picks the sound carrier out of the overall video signal by using a narrow bandpass amplifier centered at 4.5 MHz. This amplifier rejects the picture information but passes the sound information. The sound carrier is a frequency modulated signal which is converted to audio

frequencies by techniques previously discussed. Audio amplifiers boost this signal to the level that is required by the loudspeakers.

Deflection Circuits

The deflection circuits extract the horizontal and vertical synchronizing pulses off the top of the flyback pulses shown in *Figure 7-5*. These are used to synchronize the frequencies and starting points of the horizontal *(Figure 7-5b)* and vertical *(Figure 7-5c)* deflection signals produced by the horizontal and vertical oscillators respectively. The outputs of these oscillators are amplified and applied to the deflection coils to provide the proper raster scan pattern that was discussed earlier. The fast retrace of the horizontal deflection signal shown in *Figure 7-5b* is used to generate the high voltage required by the anode electrode of the picture tube *(Figure 7-4)*. A high-frequency pulse is generated by the 10-microsecond horizontal retrace and is transformed and rectified to a voltage of 18 to 25 kilovolts of dc by the high-voltage supply circuits. This high voltage is required to accelerate the electrons to provide the proper screen brightness. There also is a low-voltage power supply which provides power to all the circuits throughout the receiver.

As mentioned earlier in the discussion on frequency spectrum and resolution, some of the performance features of television depend on the channel bandwidth allotted to the television signal containing the picture and sound information. The carrier and bandwidth of these signals have been standardized by the broadcast industry and the Federal Communications Commission. Let's look at the bandwidth requirements in more detail.

WHAT ARE THE BANDWIDTH REQUIREMENTS OF RADIO AND TELEVISION?

TV Channels

Since the bandwidth of the IF amplifier in *Figure 7-12* is 6 MHz (46 - 40), then one would expect that the television channel bandwidth is 6 MHz. This is exactly correct as shown in *Figure 7-13*, which is a list of television channel frequencies. There is no channel 1, but channels 2, 3, and 4 are separated from each other in frequency by a bandwidth of 6 MHz. Channel 5 and 6 are also separated by 6 MHz but they are adjacent to the FM broadcast band from 88 to 108 MHz. Above this are 9 channels assigned to cable television (CATV channels in the midband range). Then the remaining VHF commercial broadcast television channels from 174 to 216 MHz. After this there are some additional cable channels from 216 to 295 MHz. Finally there are the 56 UHF broadcast television channels from 470 to 806 MHz. There used to be an additional 14 UHF channels extending to 890 MHz, however these have now been allocated to land mobile service. If the CATV and the broadcast channels are counted, the FCC has provided 90 channels (excluding any sub-band channels that are potentially available to CATV below the channel 2 frequency band).

Broadcast Channels		Cable Television Channels	
Channel	Frequency Range (MHz)	Channel	Frequency Range (MHz)
2	54-60	A	120-126
3	60-66	B	126-132
4	66-72	C	132-138
5	76-82	D	138-144
6	82-88	•	•
FM	88-108	•	•
7	174-180	I	168-174
8	180-186	J	216-222
•	•	K	222-228
•	•	•	•
13	210-216	R	264-270
14	470-476	S	271-277
15	476-482	T	277-283
•	•	U	283-289
69	800-806	V	289-295

Figure 7-13. *Television Channel Frequencies.*

Signal Bandwidth of TV

The feature that all of these channels share is the common 6 MHz television signal bandwidth of $f_2 - f_1$ shown in *Figure 7-14.* The picture carrier frequency is 1.25 MHz above f_1 (the lower band frequency), and the sound broadcast carrier is 5.75 MHz above f_1. Note that the sound carrier is 4.5 MHz and the chroma signal is 3.58 MHz above the picture carrier. Thus, for channel 2, which occupies the frequencies from 54 to 60 MHz, the picture carrier is at 55.25 MHz, the chroma signal is at 58.83 MHz, and the sound carrier is at 59.75 MHz, just 250 kHz below the upper frequency limit f_2 for the channel 2 band.

The picture carrier frequency is amplitude modulated by the picture and chroma information. Since the channel bandwidth is cut-off 1.25 MHz below this frequency, only the first 3/4 to 1 MHz of video information is modulated using the conventional double-side-band ($f_c - f_m$ and $f_c + f_m$) amplitude modulation. Frequencies above this in the luminance and chrominance signals are thus single-side-band modulated. As a result, the frequencies of the luminance that extend beyond 1.25 MHz from the carrier are reproduced only at 1/2 the amplitude of the signals that fit into the 1.25-MHz frequency band. The lower frequencies produce the general features of the picture; the upper frequencies the picture sharpness and detail. The lower sideband of the amplitude modulated signal from the picture carrier frequency down to the lower band limit is called the vestial sideband.

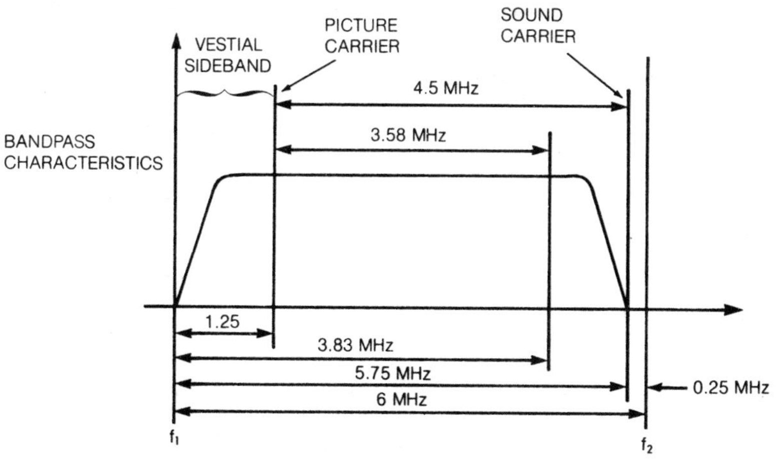

Figure 7-14. *Bandwidth Features of Commercial Television Channel.*

Sound Reproduction of FM and TV

The sound information is modulated onto the sound carrier, which is 4.5 MHz above the picture carrier, using frequency modulation with a deviation bandwidth of ±25 kilohertz. This compares with the ±75-kilohertz deviation available in FM stations. These deviations occur for 100% modulation of audio signals. Both FM and television sound can reproduce audio frequencies in the 50-Hz to 15-kHz range, though the FM signals tend to be more immune to noise with their wide deviations. In addition, FM stations tend to be multiplexed stereophonic sound and are generally reproduced with high-fidelity sound system, while the television sound is monophonic and is reproduced with what is generally classed as a poor quality system.

In the future, television receivers could certainly offer improved audio fidelity through improved audio amplifiers and high quality speaker systems. It is possible for television stations to provide stereophonic sound within the present 6-MHz bandwidth by slightly increasing the bandwidth available for the sound channel. This development, if it occurs at all, would most likely occur first in non-broadcast television networks. Certainly the cost of components both at the transmitter and receiver play an important part in this decision.

WHAT ARE SOME TYPICAL TV AND RADIO NETWORKS?

Most of the discussion thus far in this chapter has assumed that the system has been a commercial television broadcasting network, the components of which are shown in *Figure 7-15*. The studio is the central communications center for such a facility. It originates local news programs and sporting events. It receives programs from national television networks via microwave links or leased telephone lines of sufficient bandwidth to handle the television channel requirements.

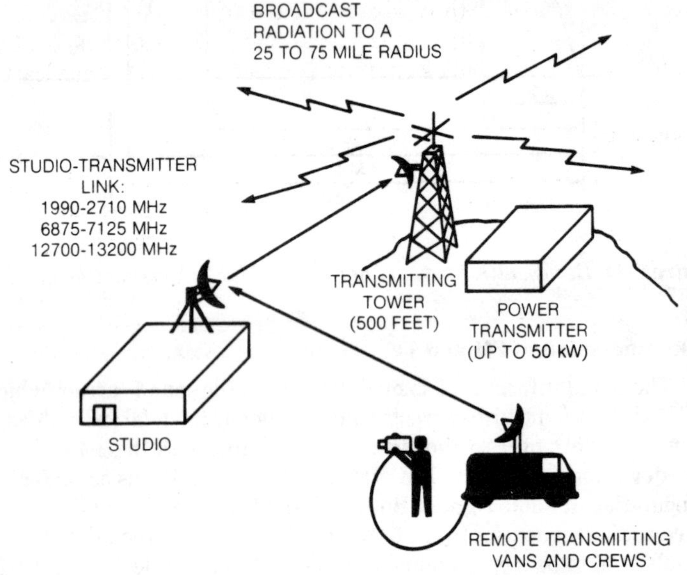

Figure 7-15. *Broadcast Television System.*

Remote Links

The transmitter for radiating the television material generated at the studio in most cases is not at the studio. The studio usually is at some convenient business location in a metropolitan area. The high-power transmitters and high antenna towers need to be located in remote areas on tops of hills or other high points overlooking the area to be served by the broadcast. For this reason, the studio signals must be relayed to the transmitting facility on a studio-to-transmitter link. The FCC lists several frequency bands for use for this purpose as indicated in *Figure 7-15*. These frequencies are used to send the television program material, communications information, and control signals to the transmitting facility. The studio antenna and the link receiving antenna at the transmitting facility must be within a line-of-sight path of each other.

Transmitter Separation

To broadcast (radiate the signal) a high-power transmitter feeds up to 50 kilowatts to a broadcast antenna on a tower about 500 feet high. Television receivers within a line-of-sight radius from the transmitting antenna of 25 to 75 miles will receive the signals. Due to this range, the FCC keeps stations broadcasting on the same channel separated by a distance of at least 200 miles (depending on whether the stations are VHF or UHF) so the stations will not interfere with each other. Stations whose channels are adjacent in frequency, such as channel 2 and 3, must be separated by about 60 miles to avoid station-to-station interference. Stations that do not share adjacent frequency bands, such as channels 4 and 5 or channels 6 and 7, may both be in the same metropolitan or broadcast area.

Radio Networks

Radio networks are similar to the situation of *Figure 7-15.* They tend to be more local and have more studio originated information for broadcast than a network TV station, but the real difference is in the program material and frequency of transmission. FM signals provide multiplexed stereo sound information in a 200-kilohertz channel in the 88- to 108-MHz range of frequencies. The transmission is line-of-sight just as in television broadcasting. The AM radio station would radiate an amplified modulated signal in the frequency range from 540 kHz to 1.6 MHz. In this frequency range a high-power (50kW) transmitted signal can travel long distances, particularly at night, because the transmission is not line-of-sight and will bounce off the ionosphere.

Radio Receivers

AM-Amplitude Modulation

The receivers for radio signals are similar to the sound portion of the television receiver. A typical AM band radio receiver structure is shown in *Figure 7-16.* As with the TV receiver, an oscillator and an RF amplifier are being multiplied together by being fed to a mixer circuit. The mixer produces the sum and difference frequencies of the two signals. In technical terms this is called heterodyning. The resultant signal, either the sum or difference, is the IF frequency. Its detector is an AM demodulator and it does have AGC circuits. The RF amplifier is tuned to provide a bandpass characteristics that will pass the signal from a single AM radio station to the mixer. The oscillator frequency is similarly varied so that the difference between the station frequency and the local oscillator frequency will be a 455-kHz IF frequency. It is this mixing and IF amplification that makes the receiver a superheterodyne structure.

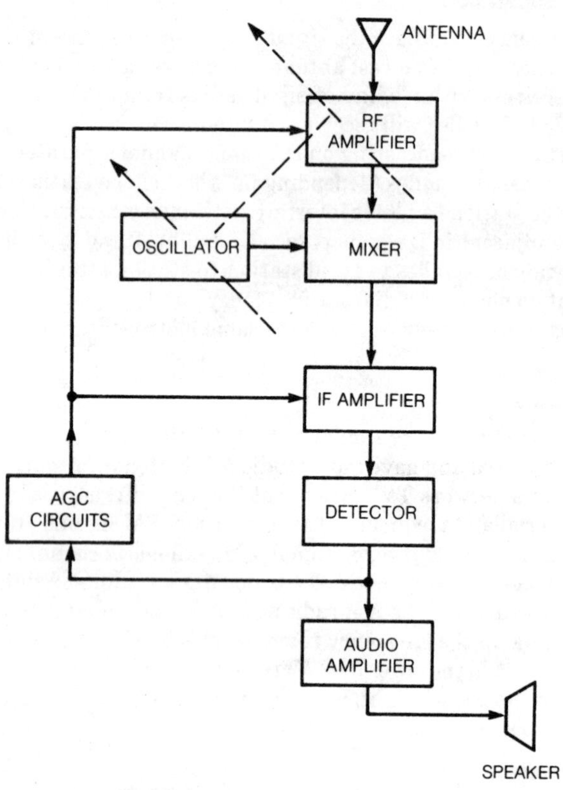

Figure 7-16. *Components of Superheterodyne AM Receiver.*

The IF amplifier boosts only the narrow range of frequencies that the signal has been converted to, blocking any extraneous signals or noise. The amplified IF signal is sent to an amplitude modulation detector which generates the audio signal and controls the AGC circuit in the same way the AGC was used in the television receiver. In the AM receiver the AGC control tends to avoid loud bursts of noise from strong stations and reduces fading and other sound level variations in the audio signal. The audio signal is amplified and sent to the loudspeaker for sound reproduction.

FM-Frequency Modulation

An FM monaural receiver would have almost an identical structure to the block diagram of the AM receiver of *Figure 7-16* except that now an FM demodulator is used for the detector. The RF and IF frequencies are much different because of the higher frequencies, but the blocks perform the same basic functions.

FM Stereo Systems

The systems for broadcasting and receiving FM Stereo are somewhat more complicated, since two separate channels of sound information must be handled. For Stereo the FM station bandwidth is allocated as shown in *Figures 7-17.* Four signals must be generated. The first signal is a composite of sound information called L + R resulting from the left channel (L) added to the sound information from the right channel (R). The third and fourth signals are the difference between the L and the R sound information. Within the bandwidth from 23 kHz to 53 kHz one of these is placed below and the other above the 38 kHz carrier frequency as shown in *Figures 7-17.* The carrier frequency in this case is unusual. It does not get transmitted. It is called a suppressed carrier. The second signal is a 19 kHz reference signal called the stereo pilot signal. It is used by the receiver to recreate the missing 38 kHz carrier signal, and to activate the stereo indicator. Also shown is an SCA signal. It is of no concern to the average FM station listener.

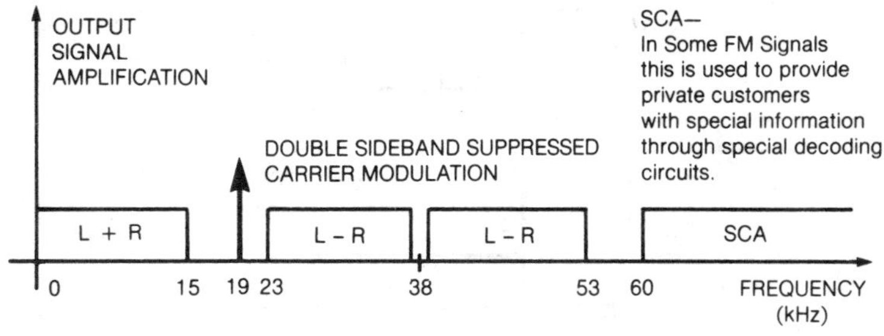

Figure 7-17. *Stereophonic FM Channel Bandwidth Allocation.*

FM Stereo Receiver

The FM receiver must separate the L + R and the L-R information and derive the L and R channel information for stereo reproduction. The basic structure of the stereo receiver is shown in *Figures 7-18.* The RF amplifier and local oscillator are tuned together, as before, so that the signal from the desired station is amplified and mixed to provide a 10.7 MHz IF frequency. This signal is boosted in level by the IF amplifiers and sent to

the FM detector (*Figure 7-18*). The simplest way to extract the L + R information is to send the composite signal through a low pass amplifier that blocks frequencies above 15 kHz. A bandpass amplifier and phase-lock-loop lock onto the 19 kHz pilot signal, double it, and provide the 38 kHz carrier. By inserting this carrier with the sideband signals, the L-R signal is obtained. Combining these two signals in a matrix circuit provides the original L and R signals. By adding the L + R to the L-R the L signal can be obtained. By subtracting these signals the R signal can be obtained. These are then sent to the L and R power amplifiers and amplified to the levels required by the L and R speakers.

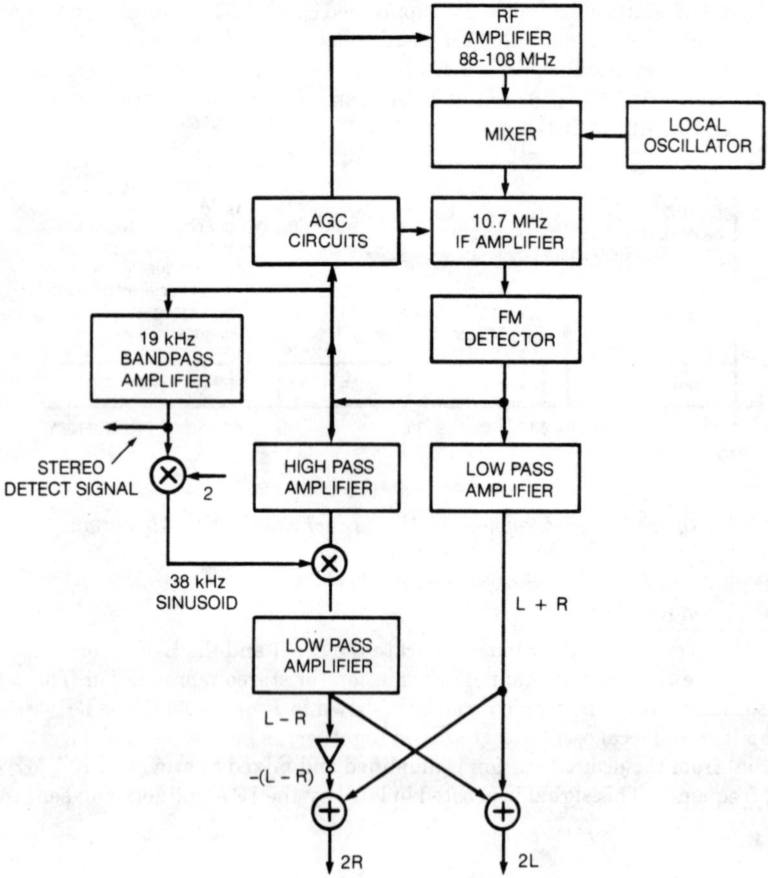

Figure 7-18. *Components of Stereo FM Receiver.*

Monaural

A monaural receiver would simply detect the stereo L + R information and deliver this sound information to a single speaker. A stereo station would send monaural information with L = R so that the L-R information would disappear and only L + R = 2L = 2R information would be received and sent to both channels of the stereo system.

Dolby Systems

Some FM stations provide a noise reduction encoding of the type used in modern tape recorders. The receiver takes advantage of this encoding to reduce the annoying effects of high frequency noise (often called "hiss" noise) in the reproduced sound.

The Dolby system uses circuits in the transmitter to compress the sound signals in certain frequency ranges prior to transmission. The receiver then expands the signals back to their original dynamic range.

For example in ordinary systems, signals varying from high to low level arrive at the receiver with the low level signals masked out by noise from the transmission. At the receiver this noticeable hiss covers up high-frequency components of the original signal.

To improve the conditions, in Dolby Systems the original signal is compressed in dynamic range at the transmitter. Therefore, the transmission noise is a smaller part compared to all levels of the sound signal. At the receiver the signal is expanded and the low levels are restored with less noise interference. The result is a sound reproduction with a lower or less noticeable hiss and better reproduction of high-frequency components of the original signal.

Radio Dispatch Network

A related system to the radio broadcast system is the radio dispatch system shown in *Figure 7-19*. A central receiver/transmitter (transceiver) station provides one-way signalling to fixed and to portable receivers to perform the function of message delivery and paging or beeper service. Two-way transceivers at fixed or portable locations provide command-response information for service vehicles such as taxis, emergency equipment, or repair vehicles. Radio telephones in cars can also be handled by two-way portable transceivers. It is conceivable that handheld transceivers can be used to send and receive data to and from computer networks in the future. Certainly, ambulances often have the capability of sending medical information ahead to hospitals while enroute to the hospital. Such a system is a type of data telemetry in use today. Stations of the type shown in *Figure 7-19* use a wide range of frequencies from 1.6 MHz to 10 GHz alloted to mobile communications by the FCC.

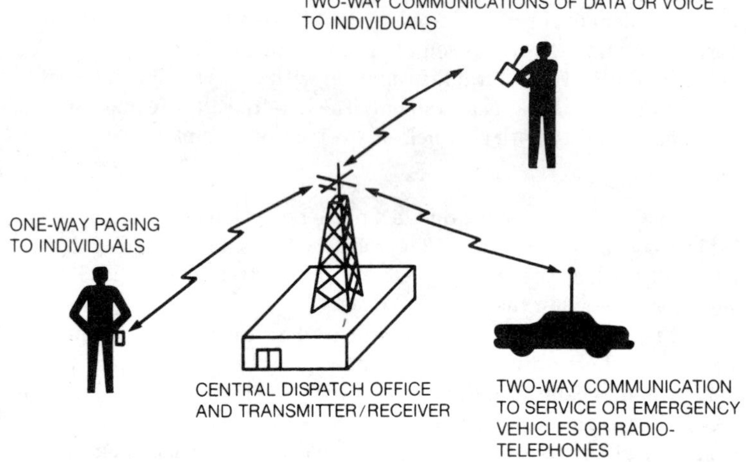

Figure 7-19. *Radio Dispatch Network.*

Cable TV

One alternative to broadcasting sound and television program material through the air with electromagnetic radiation is to use cables to distribute these signals, just as telephone lines and cables distribute private telephone conversations. The cable television system of *Figure 7-20* is an example of such an approach. The cable television or CATV is a private distribution network which sends television and other information from a central studio to subscriber's locations. The studio generates programs or leases program material from national distributors relayed to them by satellite or microwave links. The studio then sends these channels of television and sound broadcasting to the locations of the subscribers who have bought these services. The signal distribution is usually along a central trunk to which feeder lines are connected to subscriber locations. This is similar to a city water supply or other utility distribution system.

In some systems the provision is made for communications from the subscribers to the studio for service requests or information requests or subscriber polling or interaction with the cable programs. This is an advantage of the direct connected cable system. Another advantage is the low-power operation in a noise-free environment. In addition, the cable system usually can offer many channels of information all at the same time. These could include first-run movies, stock market reports, weather/news reports on a 24-hour basis, sporting events, etc. Cable could ultimately offer access to private computer and information networks, electronic game pools, burglary prevention or safety communications services, and complete home educational packages.

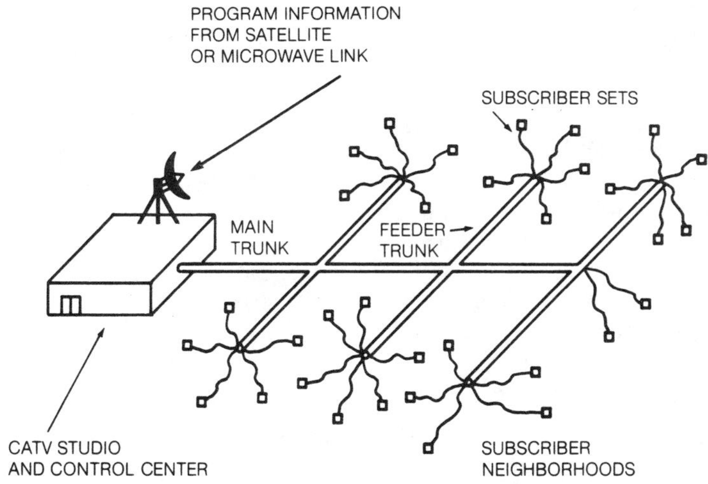

Figure 7-20. *Cable Television Network.*

DIGITAL DISTRIBUTION OF TV

All of the systems that have been discussed in this chapter have historically been analog signal systems. Television systems in the future may include more digital circuit techniques due to the trend toward digital systems for telephone and satellite communications systems. While home TV receivers may remain analog for the foreseeable future, the distribution of program material from national sources to local stations may require digital equipment. As has been discussed and will be further expanded, digital transmission of such signals offers the possibilities of more efficient and more noise-free communication of the information.

Basically, to convert a television picture to digital form requires that the video signal be sampled at twice its highest frequency and each sample converted to a PCM (pulse code modulation) equivalent. If the television signals contain frequencies up to 4.6 MHz and the picture information requires a 10-bit binary code for accurate representation, then the digital signal would have to be sent at a rate of 4.6 MHz $\times$ 2 $\times$ 10 = 92 megabits/second. This is the digital channel capacity of a T4 telephone system channel. Digital signal compression techniques can be used to send a television signal in about one-half this bit rate which would allow two television channels to be sent on a T4 line.

The more television channels that can be sent over a given telephone or satellite channel, the less expensive the "transportation" costs for the network. If television is used to produce relatively still images such as graphs or picturephone images, the bandwidth requirements can be reduced to less than 1 MHz for a bit rate of 1 MHz x 2 x 8 bits = 16 megabits/second. This is without data compression techniques. Techniques such as slow scanning and digital compression could allow such images to be sent with only 60k bits/second. Such transmissions would be quite economical because they can be handled by a single digital voice channel. However, it would have to be a very still image, and it could take several seconds for the complete image to be transmitted and formed on a receiving picture tube. Just which of these options will find sufficient commercial markets remains to be seen.

WHAT HAVE WE LEARNED?

- Television cameras convert an optical image into a series of electrical signals by scanning the picture a line at a time with an electron beam and converting each spot of the picture into an electrical voltage amplitude.
- Television picture tubes reproduce pictures from electronic signals by using the signals to vary the intensity of a scanning electron beam which in turn varies the intensity of light produced by a phosphor on the face of the electron tube.
- The horizontal and vertical scanning of a picture forms the raster of the picture tube.
- Each picture consists of a frame of 525 lines, with 262.5 lines per field and two interlaced fields per frame.
- The horizontal scanning frequency is 15,750 hertz and the vertical scanning frequency (field repetition rate) is 60 hertz.
- The television signal occupies a bandwidth of 6 megahertz.
- The color content of the picture is contained in the phase of a chrominance sine wave signal while the black-and-white content is contained in the luminance signal.

WHAT'S NEXT?

The last two chapters have dealt with the means of transmitting and receiving sound and pictorial information. A very important category of information in the modern electronic world is that of data. The next chapter will concentrate on data communications and the computer networks that utilize and process such data.

Quiz for Chapter 7

1. An optical image is converted to an electrical signal by:
   a. focusing the image on a light sensitive material and scanning the charge or conductive pattern of this material with an electron beam a line at a time.
   b. the focused image passing through a camera's transparent glass plate and interfering with the motion of a scanning electron beam so that its current varies with the image intensity and color.
   c. the optical pattern varying the conductivity or charge on the surface of a television camera which is detected by thousands of wires connected to each spot on the tube.
2. The flyback period of the electron beam scan pattern is:
   a. very short compared to the camera or picture tube optical scan.
   b. a time during which the electron beam current is ignored in a TV camera.
   c. the time during which the electron beam current is cut off in a TV picture tube.
   d. All of the above.
3. For TV scans, the horizontal scan time is ______ in comparison with the vertical scan time.
   a. long
   b. short
   c. about the same
4. Each field of a U.S. broadcast TV raster is:
   a. 525 lines.
   b. 262 ½ lines.
   c. 100 lines.
5. The vertical scan rate is:
   a. 30 hertz.
   b. 60 hertz.
   c. 15,750 hertz.
6. The resolution of a 525 line scan pattern is about the same as:
   a. 8 mm movie camera.
   b. 16 mm movie camera.
   c. 35 mm movie camera.
7. The luminance signal of a color TV video signal is related to:
   a. the color hue.
   b. the scan frequency.
   c. the color intensity.
   d. the gray level.
8. The chrominance signal of a color TV video signal is related to:
   a. the gray level information.
   b. the color saturation.
   c. the color hue.
   d. a and c above.
   e. b and c above.
9. If the chrominance is 180° out of phase with the reference signal, the color is:
   a. red.
   b. green.
   c. blue.
   d. white.
10. The chrominance signal is sent at the same time as the:
    a. horizontal blanking pulse.
    b. vertical blanking pulse.
    c. luminance signal.
11. The colors used as phosphors in the standard television picture tube are:
    a. red, green, and yellow.
    b. red, green, and blue.
    c. red, blue, and yellow.
12. In the TV bandwidth, the sound carrier is located ______ MHz above the picture carrier.
    a. 3.58
    b. 3.83
    c. 4.5
    d. 5.75
13. The sound is modulated onto the sound carrier using:
    a. AM – double sideband.
    b. AM – single sideband.
    c. AM – suppressed carrier, double sideband.
    d. narrow-band FM.
    e. wide-band FM.

14. The standard commercial television channel requires a bandwidth of:
   a. 3.5 MHz.
   b. 4.5 MHz.
   c. 5.75 MHz.
   d. 6 MHz.

15. In stereo FM broadcasts, the left and right audio information are sent by:
   a. frequency multiplexing the left channel and amplitude modulating the right channel.
   b. amplitude modulating both channels and then mixing them.
   c. providing composite L plus R and L minus R signals. The L minus R signal is amplitude modulated around a 38 kHz suppressed carrier.

16. The standard FM station occupies a bandwidth in kHz of:
   a. 15.
   b. 53.
   c. 100.
   d. 200.

17. Broadcast AM radio stations offer a sound bandwidth of:
   a. 2 to 5 kHz.
   b. 5 to 10 kHz.
   c. 10 to 15 kHz.

18. TV can be sent in digital form over a channel that has a bit rate capacity of:
   a. 4.5 megabits/second.
   b. 45 megabits/second.
   c. 92 megabits/second.
   d. 270 metabits/second.

19. If a TV receiver loses its color burst signal, the picture will:
   a. go black.
   b. go all white.
   c. lose its brightness.
   d. lose its reds and blues.
   e. lose all color.

20. The 1 MHz and above components of a TV video signal contains the following picture information:
   a. color.
   b. picture detail or sharpness.
   c. overall picture forms and outlines.
   d. All of the above.
   e. a and b above.

(Answers in back of the book)

Computer Networks and Systems

ABOUT THIS CHAPTER

In the previous two chapters the emphasis has been on methods of communications between people, either on a one-to-one basis with telegraph or telephone systems or on a one-to-many basis with radio and television broadcast systems. These have been major advances in society's ability to communicate. Another major contributor to this progress has been the advent of the modern computer. In order to utilize the power of the computer to its full potential, humans must be able to communicate with computers and computers must be able to communicate with other computers. Some of the techniques used in such communications will be covered in this chapter.

WHAT IS A COMPUTER?

Major Components

In order to understand the importance and the capabilities of computers, the operation of a computer must be understood. *Figure 8-1* shows the general components of a computer and how these components are combined to communicate with the external world. Computers consist of three major conponents: the memory, the input/output interface, and the processor. Information is handled inside the computer as coded digital signals, as explained in Chapter 2. Binary bits grouped together represent the digital codes. The meaning of a given group of bits or a particular code depends on where it occurs and how it is used in the computer. In *Figure 8-2*, the components of *Figure 8-1* are redrawn with some simple codes indicated at various points in the system. Each set of lines interconnecting components is shown as a four line group. This is called a nibble in computer jargon. Multiple pairs of wires running together like this each carrying a bit of a digital code are called buses. They may be for 8-bit bytes, 16, 32 or n-bit words.

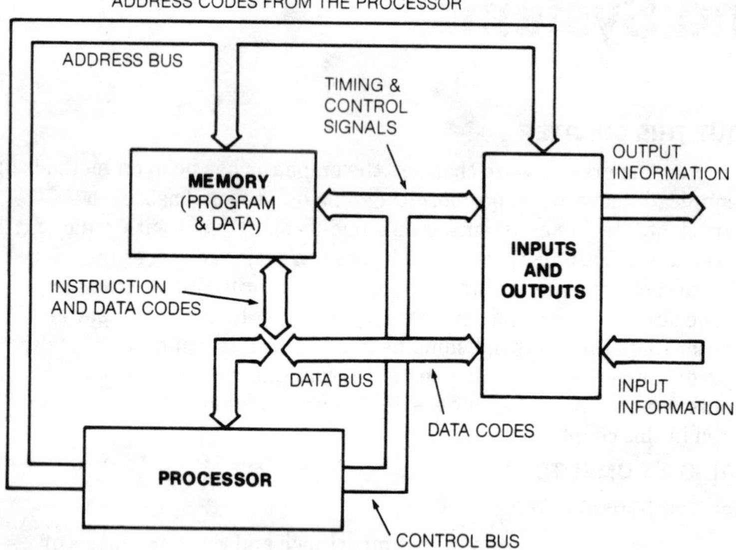

Figure 8-1. *Structure of Computer.*

Address and Data Buses

The four-line group that represents the address generated by the processor is called the address bus and contains the address code which locates particular storage cells in memory from which information is to be recalled (read) or into which information is to be stored (written into) by the processor. In *Figure 8-2,* the code 1001 sent by the processor would represent memory location 9 (see *Figure 2-6*) since the decimal equivalent of the binary number 1001 is 9. Address decoder circuits in the memory would isolate storage location 9 so that a digital code could be written into or read from it. Suppose memory control signals call for the memory to be read. *Figure 8-2* shows that the binary information in location 9 is the code 0110. This will be an instruction code that will tell the processor what to do, or it will be data information (numbers, system status, or alphanumeric character codes) that will be used by the processor as it executes an instruction.

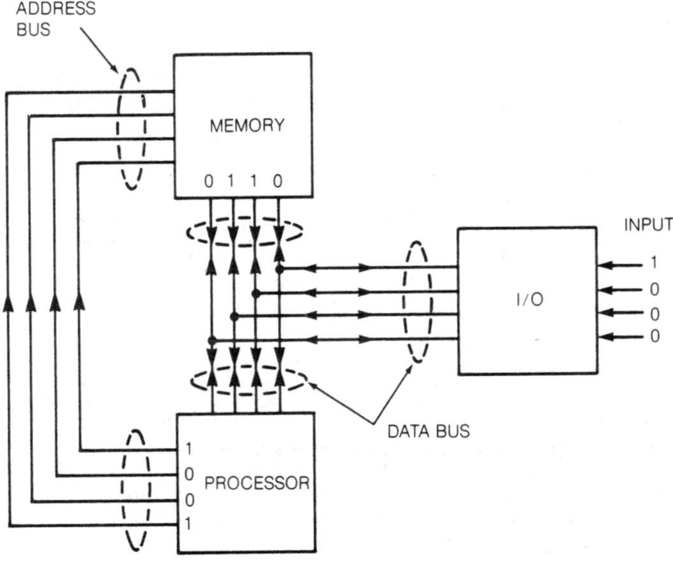

Figure 8-2. *Example 4-Bit Computer.*

Instruction Codes

If the 0110 means instruction 6 (the decimal equivalent of 0110), this code will be stored in a register in the processor, decoded and interpreted by the digital circuits in the processor. Once the processor knows that it is instruction 6, the digital circuits inside the processor will cause the events to occur that will carry out the operation indicated by this instruction code. Such events could be to move data from the input lines to the processor or from somewhere inside the processor to memory or to some output lines. The instruction could be an arithmetic operation such as addition or multiplication; it could be a comparison operation; it could be an instruction to send a character code to the output, or sense a particular input; or it could be a logical decision to determine if one numerical code is greater than or less than another. If instruction 6 means that the processor should take the data on the input lines (1000 in *Figure 8-2*) into a register in the processor, the processor, through the I/O control bus, will provide a means of switching the input lines onto the processor data lines. The processor then reads the input line information into the appropriate register inside the processor.

Overall Component Functions

From this general discussion, some of the functions of the components of *Figure 8-1* begin to become clear. The memory stores all the instructions and data needed by the system. The instructions are stored in the sequence in which they are executed. This sequence is called the computer program. It defines how the computer is to behave, and what it is to do with data that it receives from the external world, and what information it is to deliver to the external world. The input/output component connects the processor to the outside world, providing lines for information to flow into the processor from users and machines (input lines) and providing lines for information flow out to the external users and machines (output lines). The external machines can be other computers or any machine that can be controlled by electronic signals, which includes just about all machines that can be designed.

The processor acts as the central control and switching center for the computer. It performs this function by repeating the following activities over and over again:

- Generate the Address (Memory Location) of the Next Instruction
- Send the Address to Memory and Read the Memory
- Store the Instruction from Memory in the Processor
- Interpret the Instruction
- Execute the Instruction

As the processor executes each instruction in the program in the proper sequence it will behave in the manner for which it was designed. The program and the input/output interface defines the purpose of the computer.

Large Scale Computers

The variations of physical size, characteristics, and applications of modern computers is almost endless. *Figure 8-3* shows a computer center. The computer system occupies the entire room. The inputs are by typewriter and keyboard or terminals that have keyboards and a TV tube type screen. They can also be machines that provide inputs stored on magnetic tape or on punched paper cards, or bring inputs in on telephone lines or cables that are relaying information from remote locations. The outputs are special light display panels, or a hard-copy printer, or a TV tube screen, or a tape machine or magnetic disk that keeps the information in storage. The computer is a general purpose machine with a large memory. It can be a computer for solving scientific problems, a large data processing system for an insurance company, or a control center for an electronics switching system for a telephone company.

Figure 8-3. *Large-scale Computer.*
(Courtesy of IBM Corp.)

Small Size Computers

Alternatively, as shown in *Figure 8-4*, the computer could be self-contained in a small console with a television display and a keyboard terminal to provide a home computer for working on household programming and control tasks. With present solid-state technology, the computer structure of *Figure 8-1* can be placed on a single integrated circuit (called a single-chip microcomputer), as shown in *Figure 8-5*, which can be interfaced with other circuits to system inputs and outputs for a small, highly capable special purpose control system for commercial or industrial applications. With such a wide range of computer structures available, there are very few tasks that will not come under the area of computer control and communications in the future. This means that human-computer and computer-computer communications will become increasingly important.

Figure 8-4. *Computer For Use in the Home.*

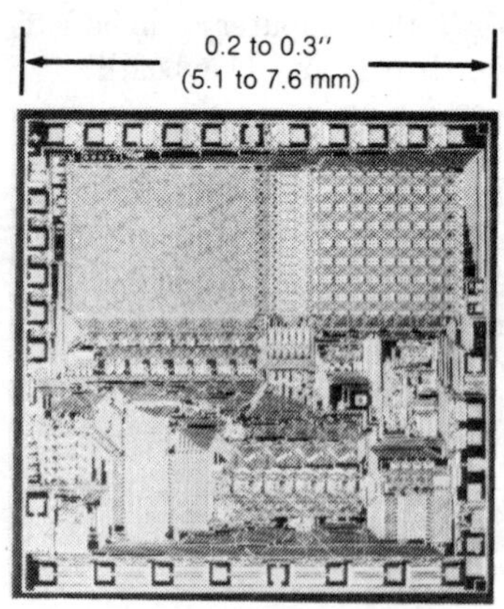

a. Integrated Circuit Chip

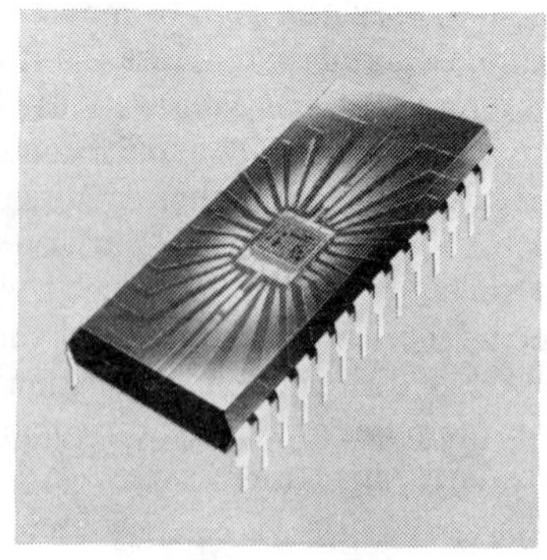

b. 28-Pin Package (may extend to 64 pins)

Figure 8-5. *Complete Computer on a Chip.*

HOW CAN HUMANS COMMUNICATE WITH COMPUTERS?

Typical Inputs

The basic component required for a human to communicate with a computer is the computer terminal illustrated in *Figure 8-6*. The terminal must provide a means for the computer to communicate either visually or through sound with the human operator, and vice versa. The most common input device is a typewriter keyboard, a calculator-type keypad, or a touch-tone telephone keypad. All of these allow a switch closure activated by finger action of the operator to be converted to a binary data code that can be received and understood by the computer. The computer can interpret such information because it has been programmed to examine the input information in the proper manner. In addition to keyboard information entry, some terminals provide joy-stick controls (like the ones used for computer games) to input information to the computer.

Much work is currently underway to provide the computer with the ability to recognize directly sounds or speech or typewritten texts, maps, photographs, and so on. Recognizing written or speech patterns would enable humans to communicate directly with machines either by writing the information or by talking to the machine. Such direct communication doesn't require typewriting skills; however, typewritten entry still offers a faster information transfer for most purposes.

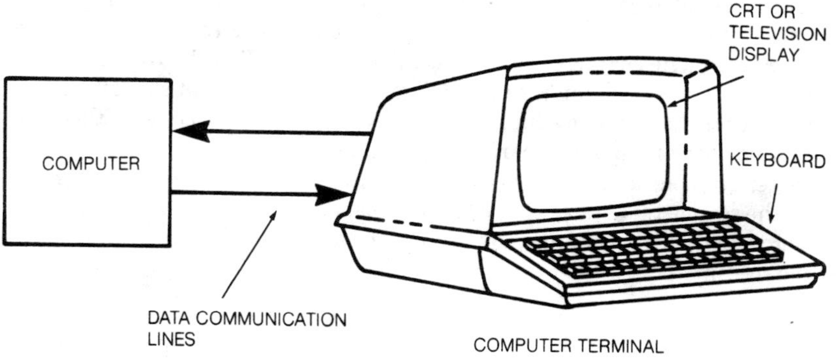

Figure 8-6. *Communicating with a Computer.*

Typical Outputs

In order for the computer to send information that is understandable to the operator, the terminal must be able to display numbers or letters on a lighted display of the type used in calculators, or to display such information on television screens, or to print such information with typewriter-like devices called printers. It is also possible to have the computer deliver the information as spoken words, though again such communication is really more limited than visual displays in terms of the amount of information that can be transferred. In all of these human-machine communication approaches, the binary codes that represent information stored in the computer must be converted to visual or sound patterns that can be recognized by the operator. The ways that these conversions can be accomplished will be considered for some of the more important techniques.

CRT Terminal Display

One of the most common output or display devices used in computer terminals is the cathode-ray tube (CRT) or television screen. In the last chapter the basic operation of television was covered and how it reproduced pictures photographed by a television camera. In the terminal, the CRT will display a page of information composed of characters and numbers or a drawing or graph organized by the computer, not an actual camera picture. To display a character, each frame is organized into a text of R rows and C columns, as shown in *Figure 8-7b*. The light emitting phosphors of a television tube can be turned on in a desired 5 by 7 matrix to display a single character on the face. The character Y is shown in *Figure 8-7a*. A sixth dot column is reserved in a column on the right side of the matrix for character spacing, and two rows of dots are reserved at the bottom of the matrix for character spacing. By energizing the desired pattern of dots, almost any character, number or symbol can be displayed on the screen. Successive characters are displayed in sequence to form words and sentences along a line across the screen and in lines (rows) down the screen.

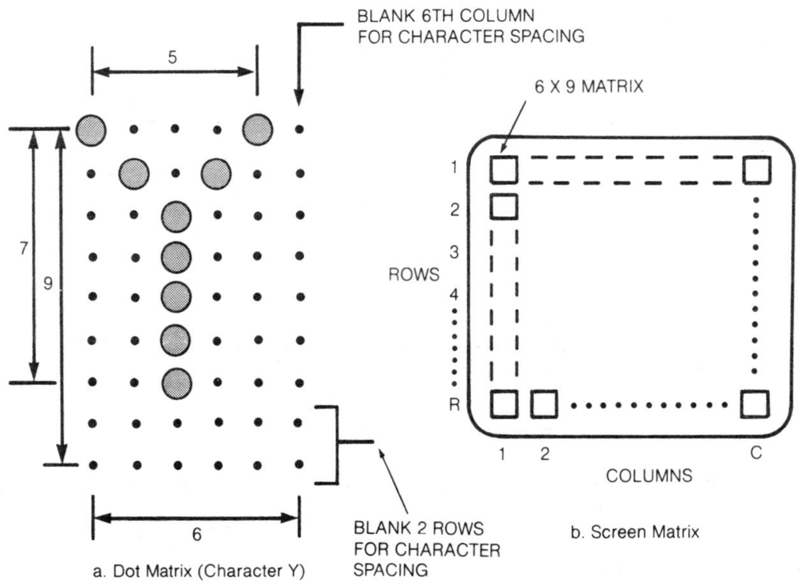

Figure 8-7. *Character Display on CRT.*

Terminal Circuitry

Inside the terminal, a semiconductor memory with the necessary bits is used to store the on-off patterns for each character and to store the character codes in the correct sequence so they are positioned correctly on the screen. The information from the memories is used with electronic circuits that generate the television raster scan pattern to provide the video signal that will display the text stored in the display memory. The basic components required are indicated in *Figure 8-8.*

The scan electronics consist mostly of counters that count the number of dots per row, the number of rows per character and the number of rows of characters. Since each character is 6 dots wide (5 display and one space), then there must be a ÷6 counter. If there are C characters/row the signal out of the ÷6 counter must be divided by C to determine the end of a row. Since there are 9 rows of dots per character (7 for display and 2 for spacing), there must be a ÷9 counter to determine when a row of characters has been displayed. Then, if there are to be R rows displayed, the output of the ÷9 counter must be divided by R to determine the end of a field. The output of the ÷6 and ÷C counter sequence will determine the horizontal retrace event; the output of the ÷R counter will define the starting of the vertical retrace event. These signals can be used to generate the pulses for blanking, retracing and synchronizing required by the television receiver electronics.

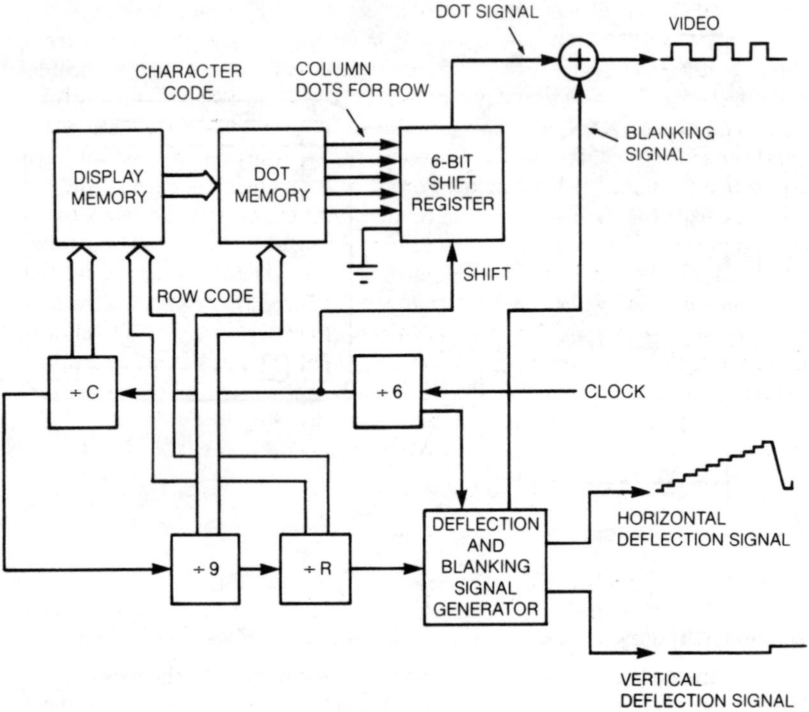

Figure 8-8. *Components of CRT Data Display.*

The video signal is a string of on and off pulses. 6C pulses occur in the horizontal line time of 53.5 microseconds. Thus, the bandwidth required of the television receiver, called a monitor in such an application, would be approximately C/9 MHz. A television monitor with a standard bandwidth of 4.5 MHz can provide a maximum of 40 characters. Modern monitors of up to 80 characters per line would require a bandwidth of 9 MHz or more.

Picture Resolution

The number of rows of characters possible depends on the number of dots used to display the character and row spacing and the number of horizontal lines in the field. If there are 260 lines in the field and 9 rows of dots per character row, approximately 30 rows would be the maximum for a TV monitor. Twenty-four rows is a typical configuration. Of course, color control signals and special deflection signals can be added to the basic circuitry of *Figure 8-8* to provide the capability for complex computer generated color pictures and graphic displays.

Speech Inputs and Outputs

Of course, visual patterns are but one way the computer can communicate with the operator. If the components of speech or words are stored in a part of the computer system, and the computer program includes techniques for assembling these sounds and words to produce meaningful sentences, the computer can output its information verbally. Such an approach is useful when the person receiving the information does not have access to a terminal with a display. For example, by using speech as an output, a user could call a telephone number that connects to an input to the computer. Commands are then entered through the touch-tone buttons which ask for information. The computer provides the information over the telephone by speaking the information. Such information could be direct answers to questions, or instructions on how to access library files to obtain the information or instructions on how to compute an answer by a certain sequence on the telephone buttons. In this way the caller and the computer could carry on a "conversation" with the computer providing the information and services required by the caller. The advantage of this approach is that it allows untrained people with a standard touch-tone telephone to access computers and make use of their services as information systems, banking services, buying services, and so on.

Hard Copy Output

There is a problem with both the speech output and the television screen output. No permanent record is generated of the communication. Often either typewritten documents or copies of television pictures are required on paper for permanent records. To provide typewritten output on paper, there must be a printer as a part of the computer terminal. To provide copies of pictorial information on paper, there must be a facsimile receiver as part of the terminal. The details on the facsimile system will be covered in Chapter 9.

Impact Printer

The printer device can take one of many different forms, depending on the speed and cost requirements of the system. The simplest form is the standard typewriter printing mechanism driven by the computer. This has been called a teletypewriter, but today is more often called an impact printer. The computer sends the teletypewriter the character codes in the sequence to be typed and the printer responds as if these keys were hit by a typist. The printing quality can be very good, but the printing rate in many cases is limited and the printing action is often noisy. This, in turn, causes operator fatigue if a terminal is operated for long periods of time.

Thermal Printer

A quieter and faster printing method with almost as good a quality as the impact typewriter is the thermal printer. The printing element is a dot pattern of thermal electrodes and the paper is heat-sensitive. By applying heat in the appropriate dot pattern (much like the pattern of *Figure 8-7a*) the characters are formed on the paper. Characters can be formed one at a time as in the typewriter or with a long thermal head a row at a time. This provides fairly high printing speeds. Some of the modern high-speed printers use an ink spraying technique instead of an impact or thermal printing technique. Others use laser beams. These offer the quiet operation with high speed and good quality characters and are used in large computer installations.

Smart Computer Terminals

An actual computer terminal may consist of several input and output options as well as providing computational power and memory. Thus, a terminal could have an ASCII (American National Standard Code for Information Interchange) keyboard for data input, a CRT display for general purpose work under quiet conditions, a printer for paper copies of printed character text, a facsimile transceiver for sending and receiving pictures, and even sound capabilities. If the terminal is itself a computer, it can perform basic computations for cash transactions and display functions while sending information for more detailed computations or for long-term storage to a central computer. The terminal is acting as a smart terminal that handles basic tasks at the terminal and uses a larger central computer for more complicated tasks. This opens the possibility of one computer communicating with another, possibly over large distances, and possibly as part of a complex computer network. Let's look at some possible network configurations.

WHAT IS A COMPUTER NETWORK?

A Network of Terminals

There are two general configurations of computer networks, and these are illustrated in *Figures 8-9* and *8-11*. *Figure 8-9* shows the connection of many relatively inexpensive terminals to one large central computer. This allows many users to rent only that portion of the central computer that is needed only for as long as it is needed. The users are time-sharing the computer. Thus, ultimately, the costs of the central computer and all the things the central computer can do are shared by all of the terminal users. The structure of *Figure 8-9* need not be for remote locations of many different organizations. The terminals can all be in the same organization. This allows the computing power to be delivered throughout a company to the desks and factory stations or wherever it is needed.

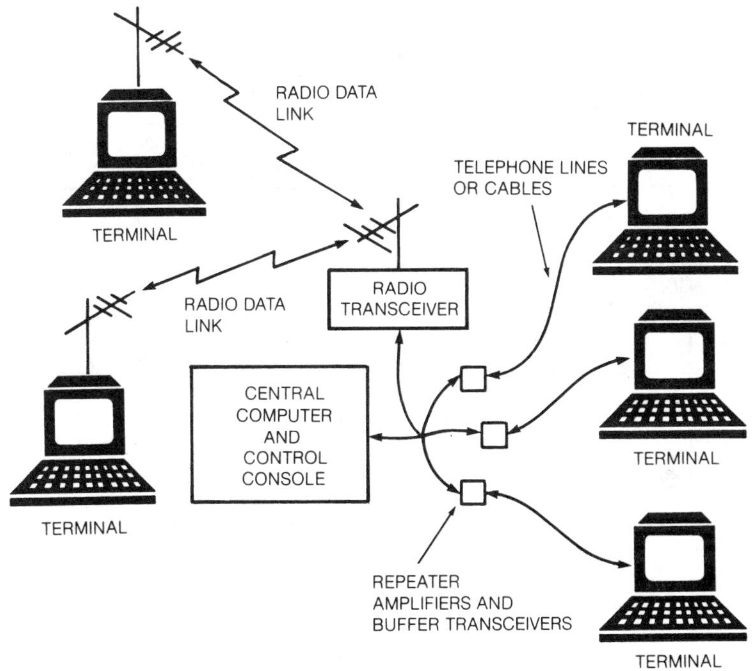

Figure 8-9. *Time-Shared Computer Network.*

The data is carried electronically over telephone wires or through space using radio waves. Most users never even see the main computer that they are using. In fact, the computer could be located in another state or country with the data communication link occurring over long-distance telephone lines, cables, microwave radio links, or via satellite data links.

Processor Activities

Since in the situation of *Figure 8-9*, many users may be sending data and programs to the computer at the same time, the computer must keep track of which user it is responding to at any given time. It also must know where it left off in the overall job being executed for a particular user as the users are given access to the computer on a time-shared or time-multiplexed basis. The computer keeps track of which block of memory and which program in that memory is being used by each terminal. It controls the time sequence in which the computer will respond to the terminals and the telephone or radio connection to the terminals. It is possible for many terminals to time-share the central computer because most terminals communicate information to and from the computer in very short bursts. Often there are long delays between these data bursts as shown in *Figure 8-10*.

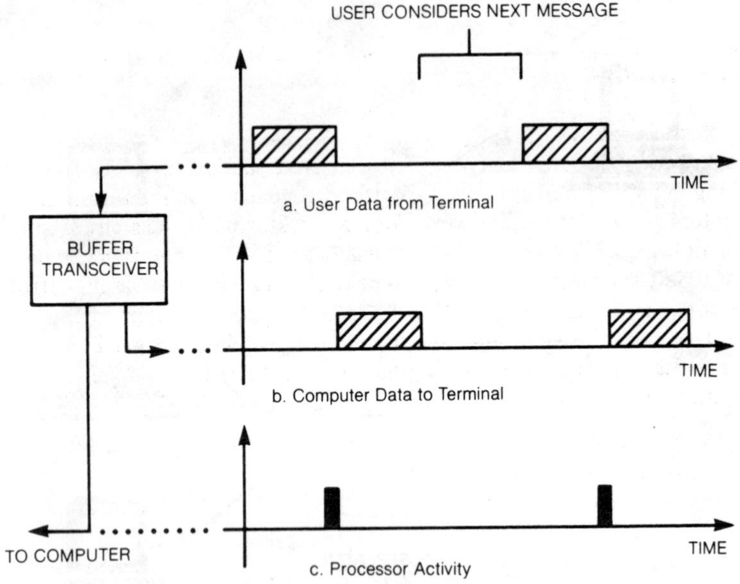

Figure 8-10. *Burst Data Nature of Computer Communications.*

For example, if a given terminal sends a 10 character command or data sequence to the computer at 30 characters per second, the transmission over the telephone line to storage locations in a buffer transceiver at the computer would take about one-third of a second. The computer can receive such information from the buffer, respond to it, and send it back to the buffer in less than a thousandth of a second. The buffer then transmits the response information over the telephone line in from one-tenth to one-half second, depending on the length of the response message. The terminal user sees an almost immediate response to a command, at which time the next response or command is considered, which could take seconds to minutes of thinking time. Even if the user immediately responded with a message, it would only be one-third second before the computer would deal with the input. Thus, the computer rarely has to spend more than a thousandth of a second per second per user, which means it could handle a thousand terminals with no more than a second delay in response to any given terminal. In this situation, the buffers transceivers are performing a significant function. They handle the routine tasks of transmitting and receiving to the terminals at the low data rates over the telephone lines or radio links. These techniques will be discussed later in the chapter.

Distributed Computer Networks

The other general structure for computer networks is shown in *Figure 8-11.* This structure includes many computers throughout the network. One version of this structure uses the central computer as a large scale unit with each of the satellite computers as small special task computers. Systems that are set up this way include transaction or point-of-sales terminals for large merchandising chains, criminal records networks, banking networks, and automated factories. In each case, the terminal computers accumulate information for a given station, do local processing and maintain local control and file information. This information can be requested by the main computer, or sent to the main computer for central processing, analysis, interpretation, filing, and so on. Managers at the central computer location can issue commands to update financial or inventory files, gather billing data, change manufacturing control procedures, and so on.

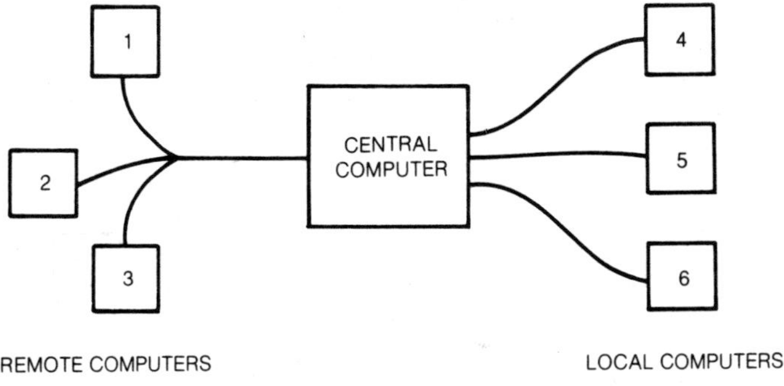

Figure 8-11. *Centralized Control Distributed Computer Network.*

The main difference between this system and that of *Figure 8-9* is that the terminals are themselves computers, and the computational and processing power of the system is distributed throughout the system, with the central computer still maintaining overall control and providing large computational capacity and memory for use by the satellite or terminal processors. Of course, all of these processing elements could be in the same building, forming one overall computer facility with multi-processing capability. This is the structure used by some of the family of large general purpose computer installations. Many processors are dedicated to handling the input/output functions and a central processing unit handles the allocation of computer and memory resources. Alternatively, the system of *Figure 8-11* could be spread out over thousands of miles with telephones, microwave, and satellite links providing the connections.

Distributed Computers of the Same Size

Figure 8-12 illustrates a different type of distributed computer network which has comparable size computers distributed over an area and communicating with each other. The central office computers of telephone electronic switching centers form a network similar to this structure. If information is to be sent from computer 1 to computer 5, there are several paths available. Data could be sent over the path from 1 to 2 to 5, or the path 1 to 6 to 5, or the path 1 to 3 to 4 to 5. For any of these paths, the intermediate computers (2, 3, 4, or 6) are acting as message relay devices. For the system to operate reliably, each computer in the system must know the source and destination of the message. Thus, each message must contain information to indicate which computer in the network is to receive the transmission. This means that each computer in the system is assigned an identification code or address code. When one computer wants to transmit data to and receive data from another computer, it includes location codes at the first of the transmission. This is similar to the address of the sender and receiver on a letter that is sent from one mailbox to a post office to another post office and finally to a destination mailbox.

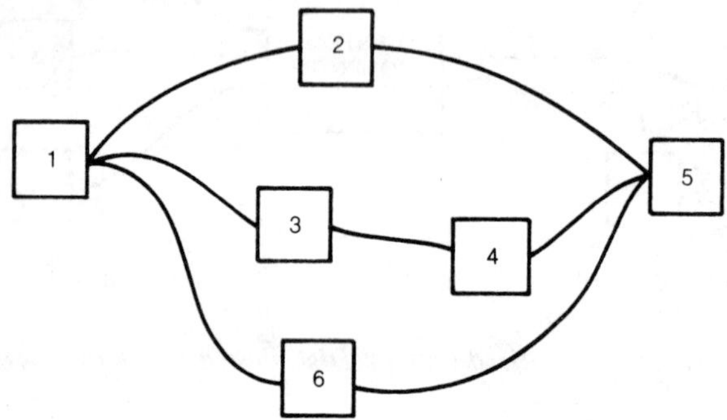

Figure 8-12. *Distributed Computer Network.*

In the computer network, each relay computer checks the destination code. If the code is the same as the computer's address, the computer receives the message, processes it, and responds with any needed return message to the source computer. If the destination code is for another computer in the network, it determines an available path for transmission to the destination computer, either directly or through another relay computer, and sends the message on to its final destination. This entire process of enabling one computer in a distributed network to send information or route information to another computer utilizes the technique called packet switching.

Packet Switching

The name packet switching derives from the fact that computer communications are sent in fixed length data bursts called packets. If a long message is to be sent, it is first broken down into packet length portions and sent a packet at a time. The common structure of a packet is indicated in *Figure 8-13.* It usually consists of around 128 bytes (or $128 \times 8 = 1024$ bits), with the first part of the packet providing routing and distribution information and the last part of the packet containing the data being transmitted. In the control part of the packet, the source and destination codes, the packet number or termination code, and the message priority information is sent. A relay computer (one whose address is not the same as the destination address) only has to look at these control bytes and send the entire packet on its way. The destination computer strips the message from the control bits, and on the basis of the control bits stores this portion of the message in the memory with the other portions that have already been received. If this is the last packet in the overall message, the destination computer starts processing the information and generating any response message.

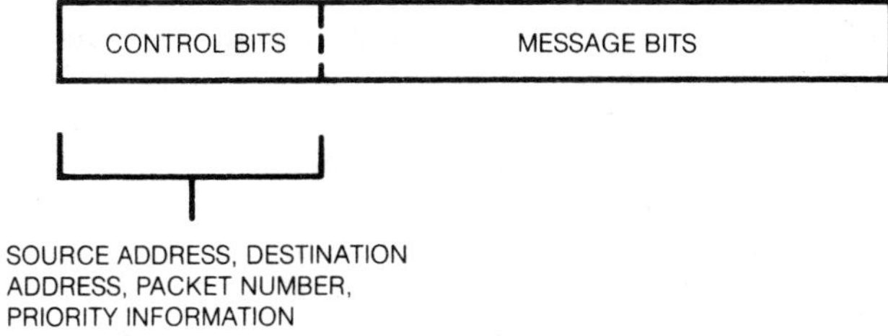

Figure 8-13. *Structure of Data Packet.*

The overall concept of packet switching is not complicated since the overall system is acting like an electronic postal system. It is an extremely useful technique since it allows a source data station to send data only to a selected terminal in a network linked by radio or satellite or telephone links.

Whether a computer network is time-shared or distributed or is just a single computer with a remote data terminal, the data communication is usually carried on a single line or a single radio frequency carrier a bit at a time. As the data is transmitted, it is subject to two conditions. The first is noise which may prevent the signal from arriving without error. The second is security. Data thieves may be attempting to intercept and receive the data and use it for their own purposes. Modern data communications networks must be aware of these conditions and take precautions against them.

HOW IS DATA COMMUNICATION CONTROLLED?

If data is to be successfully and securely sent from one location or terminal to another, several factors must be controlled. First, the basic data transmission format used to send the data over a single wire or carrier frequency must be agreed on between the transmitter and the receiver. Second, transmitter and receiver must agree on procedures to detect any error that has occurred in the data during transmission due to noise or equipment failure, and what is to be done if such an error is detected. Third, they must agree on any coding that is to be used to make the data private or meaningful only to those that know the coding procedure. Fourth, if techniques for data or bandwidth compression are used to reduce transmission costs, both transmitter and receiver must use the same compression procedures in order for the transmission to make sense to the receiver. Let's discuss each of these further.

Data Format

Data that is sent a bit at a time along a single wire or radio frequency carrier is called serial data. There are two basic serial data formats used: synchronous and asynchronous data transmission. The asynchronous approach is used for sending 8 bits (a byte) of data at a time as shown in *Figure 8-14*. Usually it is used in communicating from a low cost terminal to a central computer over telephone lines or to send data from a small or home computer to its input and output devices. In this type of transmission, the line is held at a 1 (a high-voltage level) until data is to be sent. When the sender begins to transmit the line is brought low, the receiver detects this transition and waits for a time T/2 to check to see if the line is still low. This is called the start bit and should be low if transmission of data is occurring. Satisfied that the beginning transition is a start bit, the receiver checks the line every T seconds and inputs the 1 or 0 (low-voltage level) on the line as the next bit. After the last bit of the number of bits in the data group (usually fixed by the transmitter and receiver at 5 to 8) has been received, the receiver checks the line at least once more T seconds later to see if the line is a 1. It should be, because this is the stop bit, and it being 1 is another indication that the transmission was successful.

Since a typical pattern is 1 start bit, 1 stop bit, and 8 data bits, with the 8 data bits representing an ASCII character code, it takes 10 bits to send each character. If 10 characters are to be sent per second, the bits would have to be sent at a rate of 100 bits/second, called 100 Baud. T would be 0.01 seconds for this case. Typical telephone line Baud rates are 110 Baud and 300 Baud. If the standard telephone line is used, the communication is usually done using a frequency-shift-keying (FSK) modulation procedure. The device that handles the modulating-demodulating function is called a modem.

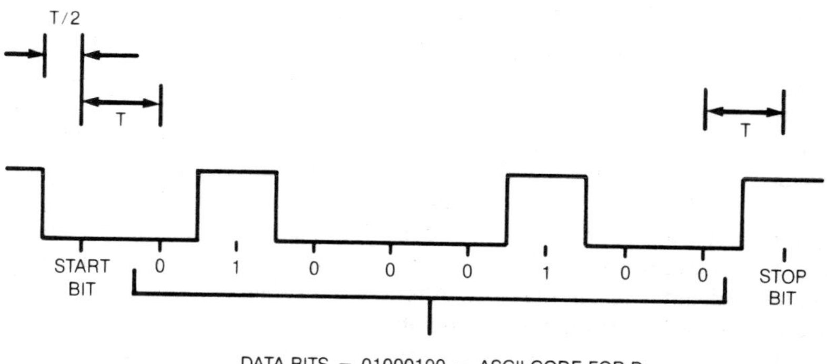

Figure 8-14. *Asynchronous Serial Data Format.*

Modems

The modem and an acoustical coupler can be used to acoustically couple the tones that represent ones and zeros of the digital transmission to the telephone speaker and microphone. Additional functions can be designed into the modem. It can be used to automatically dial a remote computer, answer the call from another computer and hang-up after a digital "conversation." As shown in *Figure 8-15*, the modem transmits a 1070 hertz tone for a 0 and a 1270-hertz tone for a 1. On receipt, it recognizes a 2025 hertz tone for a 0 and a 2225-hertz tone as a 1. These frequencies are for the source modem originating the transmission. The frequency assignments are reversed for a destination modem receiving the information.

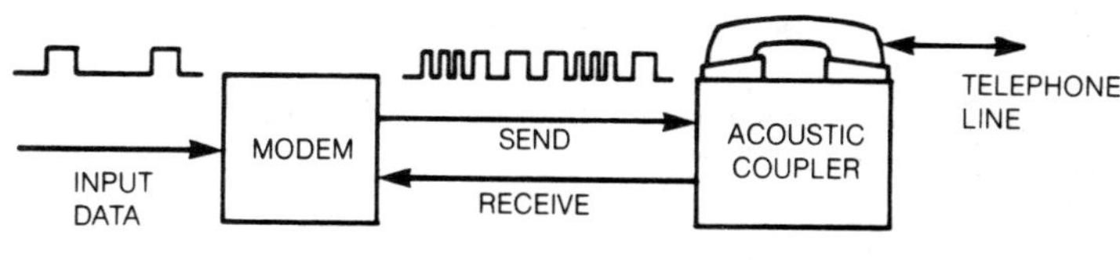

	Modem Transmitting Tone (Hertz)		**Modem Receiving Tone (Hertz)**	
	0	1	0	1
Modem Called Computer	1070	1270	2025	2225
Computer Called Modem	2025	2225	1070	1270

Figure 8-15. *FSK Transmission of Data over Telephone Line.*

Synchronous Data Format

The synchronous format is useful when a long string of bits are to be transmitted. To verify that a data transmission is coming to a particular receiver, a synchronous data receiver checks incoming bits one after another as they are received until a match is made between a received 8-bit group and a reference code. After the first match, the receiver again checks the next 8-bit group versus a reference code to make sure transmission really is occurring. Once the receiver is satisfied that the synchronization characters for the reference code have been received, it will accumulate the bits that come after as the data. The process will terminate when the receiver recognizes a code for the end of message, or when a preselected number of bytes have been received, depending on the agreed formats. The transmitter and receiver must both use the same start synchronizing characters, stop codes and/or procedure, and bit rate.

Error Checking

If the data is to be useful it must be sent without error. This is not always possible since system noise may change a 0 to a 1 or vice versa. Further, the receiver may miss part of a transmission. The data communications receiver must check to see if an error has occurred and either correct the error or, if this is not possible, ask for a retransmission of the data. Many error detection and correction schemes depend on a parity check. For example, if the transmitter always sends an odd number of ones, the receiver can check for this condition called odd parity, and detect an error if there are an even number of ones. This and most error detection schemes assume that it is very unlikely that more than one bit will be in error in any given transmission. Such a one-bit error would change an odd parity code to even parity, and would be recognized easily as an erroneous data code. By going further and adding parity bits in selected positions that check parity over different parts of the transmitted data, it is possible to even indicate which bit is in error and correct it, avoiding the time consuming approach of a retransmission.

Hamming Code

One such code is the Hamming Code. It is illustrated in b of *Figure 8-16* for a 7-bit data code. Assume that the true data code is 1000100 as shown in a (which is the code for ASCII letter D, *Figure 8-14*). The Hamming Code is a 10-bit code with parity bits P_1, P_2, and P_3, added in the bit positions ahead of X_1, in between X_1 and X_2 and between X_3 and X_4 as shown in b. P_1 is chosen to make the parity of P_1 taken with every other data bit to be 0 or even. That is, if the combination of bits in X_1, X_3, X_5 and X_7 is such that the number of 1's present is an even number, then P_1 will be set to a 0 so that the parity of the combination including P_1 will be even. If the X_1, X_3, X_5 and X_7 combination of 1's is odd, then P_1 will be set to a 1 so that the combination parity including P_1 will be even.

a. DATA TO BE SENT $x_1x_2x_3x_4x_5x_6x_7$ 1000100

b. TRANSMITTED DATA $P_1x_1P_2x_2x_3P_3x_4x_5x_6x_7$ **01**0**0**01**0**100

c. DATA GROUPS WITH EVEN PARITY:

$P_1x_1x_3x_5x_7$	**0**1010 (2 ones)
$P_2x_2x_3x_6x_7$	**0**0000 (0 ones)
$P_3x_4x_5x_6x_7$	**1**0100 (2 ones)

d. RECEIVED DATA WITH ERROR IN BIT x_6: 01000101**1**0

e. P's COMPUTED FROM RECEIVED DATA:	$P_3 = 0$	$P_2 = 1$	$P_1 = 0$
f. P's RECEIVED WITH DATA:	$P_3 = 1$	$P_2 = 0$	$P_1 = 0$
g. CHANGES IN P's:	$C_3 = 1$	$C_2 = 1$	$C_1 = 0$

h. DECIMAL EQUIVALENT OF $C_3C_2C_1$ BINARY CODE: $(110)_2 = 6$

i. THEREFORE BIT 6 IS INCORRECT — RECEIVED A 1, CHANGE TO A 0
DATA CODE TRANSMITTED SHOULD BE: 1000100

Figure 8-16. *Use of Hamming Code to Detect and Correct Data Errors.*

Similarly, P_2 is chosen to provide even parity when parity for P_2 and every other pair of data bits is computed – the combination P_2 X_2 X_3 X_6 X_7. Finally, P_3 is chosen to make even parity of the P_3 through X_7 bits – the combination P_3 X_4 X_5 X_6 X_7. As shown in c of *Figure 8-16*, P_1, P_2 and P_3 have been computed for the letter D at the transmitter and inserted before transmission as 0, 0, 1, respectively.

The received code is shown in d of *Figure 8-16*. It has an error in bit X_6. From the received data, P_1 is computed as 0, P_2 as 1 and P_3 as 0 (e of *Figure 8-16)*. These are compared to $P_1 = 0$, $P_2 = 0$, and $P_3 = 1$ received with the data (f of *Figure 8-16)*. Any change between the received P's and the ones computed from the received data are recognized and set C_1, C_2, and C_3 equal to 1, depending on the respective combination of P's that produced the error. As shown in g of *Figure 8-16*, $C_1 = 0$, $C_2 = 1$, $C_3 = 1$ because of the changes in the P_2 and P_3 combinations. The decimal equivalent of the binary code $C_3C_2C_1$ indicates which bit is in error. For example, because bit 6 is corrupted (which would indicate the letter F is received instead of the letter D), C_3 and C_2 were set equal to 1 indicating data bit 110 or bit 6 is in error. As shown in i of *Figure 8-16*, correcting X_6 from a 1 to a 0 reconstructs the transmitted code for D.

There are many error detecting and error correcting procedures that are used in data communications. The Hamming code is one of the simpler and more efficient of these approaches and it serves to illustrate the basic concept of such codes.

Data Security

Another problem involved in some data communications such as financial transactions or military operations is data security. The way to keep data thieves from stealing or misusing such information is to send the data in a secret code, just as spies have done throughout history. Codes can be as simple as rotating the alphabet by a fixed displacement to an encoding procedure that changes the meaning of each code symbol on a random basis for each character transmitted. Whatever the coding procedure, both the transmitter and receiver must know it and must have the proper encoding and decoding circuits. The basic system components are shown in *Figure 8-17*. The message goes through a device (encryptor) that scrambles or encodes it according to some encryption technique which is dependent on a key code. At the receiver, the reverse of this scrambling process is performed, using the same key to unlock the code. The decryptor will then deliver the transmitted message in a correct and understandable form. Any receiver not equipped with the proper decryption circuitry and not in possession of the key code will not be able to understand or use the data.

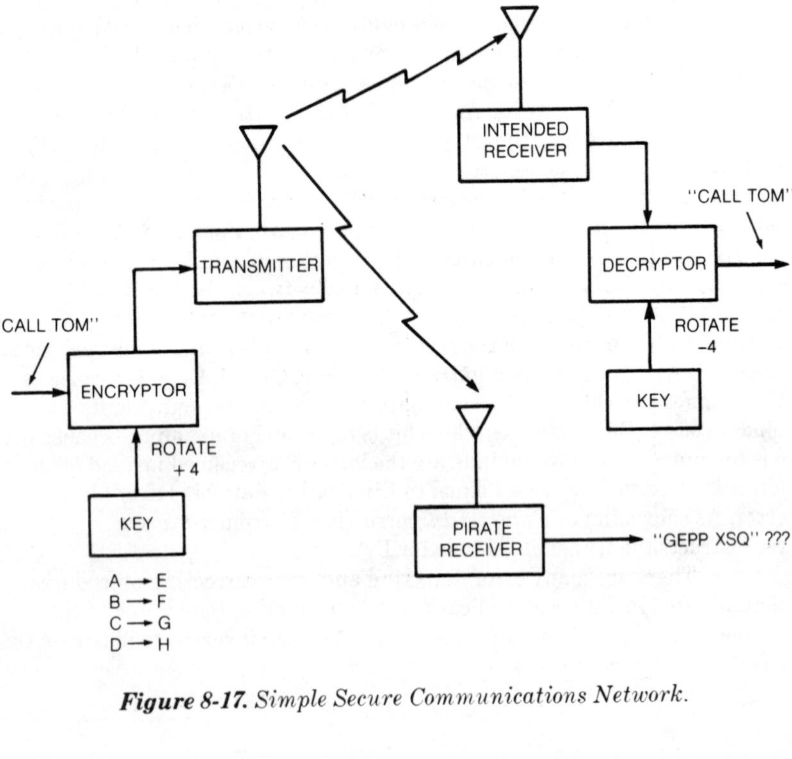

INTENDED RECEIVER
"CALL TOM"
TRANSMITTER
DECRYPTOR
"CALL TOM"
ROTATE –4
ENCRYPTOR
KEY
ROTATE + 4
KEY
PIRATE RECEIVER
"GEPP XSQ" ???
A → E
B → F
C → G
D → H

Figure 8-17. *Simple Secure Communications Network.*

A simple example of a code that could easily be broken if parts of the message or the message context were known is to replace each character by the corresponding character 4 places down. In this approach, the message "Call Tom" would become "gepp xsq" which is what a potential data pirate would receive, unless he had the rotate 4 decryption procedure. Of course, the more important the data is, the more complex and expensive the equipment to provide absolute security.

Data Compression

If the data transmission is not strictly computer data but still is in digital form, considerable bit rate and thus transmission cost savings can be obtained using data compression techniques. *Figure 8-18* illustrates one simple technique known as differential PCM or DPCM coding. The waveforms simulate a signal for a given horizontal line of a television or picturephone or facsimile video signal for two successive frames. These signals have a wide dynamic range and would require 8 to 10 bits of PCM data for accurate transmission. Such transmission would require about 90 megabits per second, which requires one-third of a T4 telephone digital channel. Thus, only three television broadcasts could be sent on a single T4 channel, each paying one-third of the lease costs of the channel.

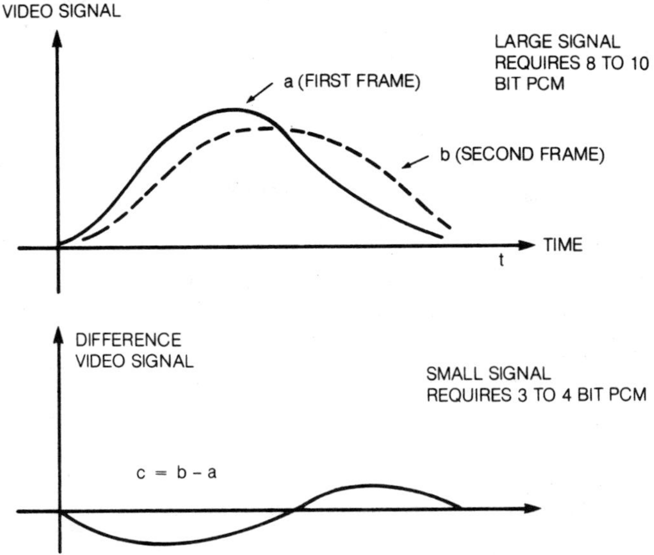

Figure 8-18. *Video and Differential Video Signals.*

Difference Signal

Instead of sending the actual signal, the difference between successive frame video signals for each line is sent, this difference signal can be sent accurately with 3 to 4-bit PCM. This means that 6 television channels could share a T4 line or one channel could be sent on a 45-megabit/second T3 line. In other words, the transmission costs could be halved, making a significant impact on operating costs.

System Components

Figure 8-19 shows the basic components required to make use of DPCM techniques. The first frame video signal, a, is stored in digital form both at the transmitter and receiver. It is converted back to an analog signal and subtracted from the new video being received for frame b. This results in the low amplitude video signal, c, that is transmitted using 4-bit PCM. At the receiver, the frame a video for the line, converted back to analog form, is added to the analog equivalent of the DCPM signal, c = b - a, to get the new b video for the line. The b video in turn is saved for reconstructing the next line. The original TV picture is thus reconstructed. Additional circuitry may have to be added to avoid problems caused by rapid motion in the picture, but the transmission can still be achieved with a 45-megabit/second channel. The particular compression technique of *Figure 8-18* and *8-19* is called inter-frame DPCM. There are many more complicated procedures which can reduce the bit rate requirements even further.

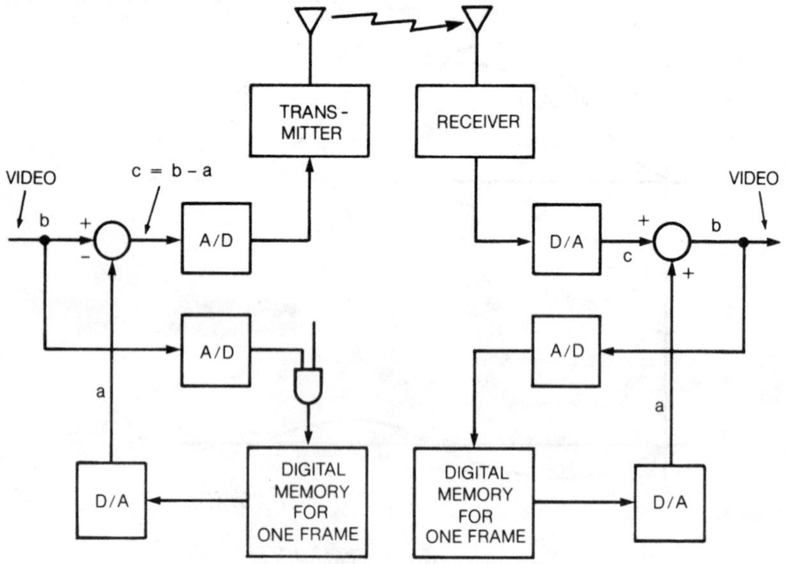

Figure 8-19. *DPCM Data Compression System Components.*

Because of the versatility and precision of digital data transmission and digital computation, communications of all sorts will rely even more heavily in the future on computer networks and computer control of digital communications. The common functions among systems; the superior noise, speed, and error performance; the high reliability, low power, and high complexity; the availability of large system functions in a small size due to integrated circuits makes this conversion to digital almost inevitable.

WHAT HAVE WE LEARNED?

- A computer is a system composed of memory, a processor, and input-output capability. The processor controls the system through a program of instructions stored in the computer memory.
- The computer program and the input and output interface defines the operation of the system.
- Computers come in all sizes from a single circuit package to large rooms full of equipment.
- A terminal is used to communicate with a computer. It provides typewritten or television screen image displays of computer outputs and a keyboard for sending information to the computer.
- The terminal could itself be another computer, in which case the user communicates with his terminal computer for basic tasks and the terminal computer communicates with the larger computer for more complicated tasks.
- Computer networks can be a time-shared central computer with many operating terminals, which may or may not have computing capability, or it can be a distributed computer system.
- Packet switching is a technique that allows computers to route information through a computer network much like a post office routes mail.
- Error correction and detection techniques provide for error-free detection even in the presence of noise.
- Data communications can be made secure as they are transmitted from one location to another by special coding techniques, the key to which is understood by both the transmitter and receiver.
- Data compression techniques can save significant transmission costs.

WHAT'S NEXT?

In the next chapter, the production of pictures and written text on paper through the use of facsimile systems will be explored. Facsimile systems share many of the features of television and picture telephone communication systems. The added features that they offer is production of a permanent record of these types of visual information on paper.

Quiz for Chapter 8

1. A computer program is:
   a. a sequence of instructions that tells the computer what to do.
   b. a sequence of binary codes stored in memory that causes the computer to behave in a prescribed manner.
   c. a plan of activity for the machine.
   d. all of the above.
2. Computers consist of the following components:
   a. memory, processor, and human operator.
   b. processor, data and program memory, and I/O.
   c. memory and I/O devices only.
3. The processor in a computer provides the following:
   a. It stores the computer program.
   b. It provides the input/output function.
   c. It acts as the central control element, interpreting and executing the program instructions in the proper sequence.
4. A computer terminal provides:
   a. a means for humans to exchange information with a computer.
   b. a connection of wires used to interconnect computer components.
   c. a sequence with which an input to a computer is terminated.
5. Computers communicate with humans by using:
   a. optical inputs.
   b. switch inputs.
   c. optical displays.
   d. printed outputs.
   e. All of the above.
6. A CRT display that provides 80 characters per line would require a video bandwidth of approximately:
   a. 4.5 MHz.
   b. 9 MHz.
   c. 12 MHz.
7. In a computer network involving human I/O the great majority of time is spent:
   a. by operators sending information to the computer and vice versa.
   b. by computers sending the information to other computers.
   c. by computers providing the computations and processing required.
   d. a and b above.
8. In a time-shared computer network:
   a. many terminals are connected at the same time to a central computer.
   b. information is time-multiplexed into and out of the computer.
   c. the information is sent to the computer over a time-multiplexed telephone line.
   d. a and b above.
9. A distributed computer network:
   a. has multiple computers and terminals in the system.
   b. has a large central computer with many small computer terminals.
   c. distributes the computing tasks throughout the system.
   d. a and c above.
   e. b and c above.
10. A data packet consists of:
    a. a single byte of data.
    b. a long string of data bytes.
    c. a long string of data bytes with source and destination codes.

(Answers in back of the book)

Facsimile Systems

ABOUT THIS CHAPTER

All of the communications techniques examined thus far in this book have dealt with the transmission of voice signals, music and sound, visual information, data, and facts. Only in the case of certain news and data communications has there been a permanent copy of the results of the transmission. In this situation the copy is typewritten output, originally generated by a typewriter keyboard or by a computer printer. This chapter deals with techniques that transmit the contents of source documents that may be pictures, printed pages, typewritten pages, or film to a remote location where a copy of the document or film is reproduced on paper or photographic film. The process is called facsimile.

WHAT IS FACSIMILE?

The dictionary defines facsimile as an exact copy or as the process of transmitting printed matter or still pictures by a system of either telephones or telegraph or radio for reproduction. The key words in this definition are copy and reproduction. Everyone is familiar with copy machines that accept an original document containing printed and pictorial information and then duplicate this visual pattern on another piece of paper. Facsimile systems do exactly this. The difference between a facsimile system and an office copier is primarily in the location of the original and the copy. In the facsimile system, these are located remotely some distance from each other, while in the office copier they are located in the same machine in the office. Thus, the basic facsimile system is the form of *Figure 9-1*. The document or photograph (source document) to be transmitted is input through the sending unit of the facsimile system. This unit performs the conversion from the visual information on the document or film into electrical signals representing the information. These electrical signals modulate a transmitter so that the information can be sent to the receiver electrically through the transmission links discussed – wires or cables and amplifiers, electromagnetic waves in waveguides or electromagnetic waves through space. The receiver first converts the transmitted information back to the same sort of electrical signals generated by the sending unit, and then converts the patterns of the electrical signals to a copy of the original source document, whether picture or printed page.

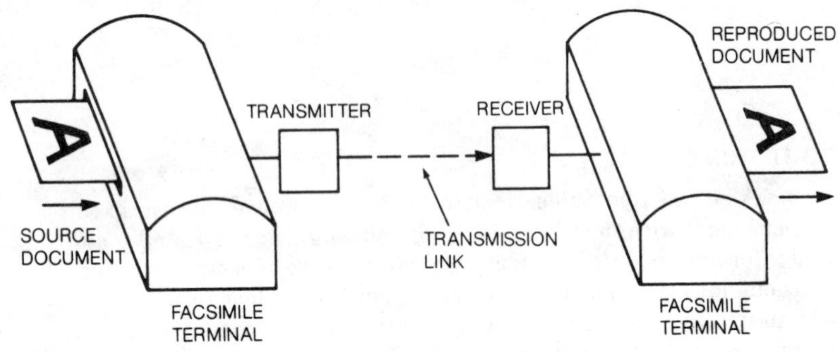

Figure 9-1. *Basic Facsimile System.*

A slight variation of the basic facsimile system is shown in *Figure 9-2*. The sending unit in *Figure 9-2* is an electronic sketchpad that accepts handwriting as an input. A facsimile receiving unit generates the output. The pen positions on the sketchpad are encoded as a sequence of electrical signals which when decoded by the receiving unit will duplicate the handwritten sketch or message composed at the transmitting unit. This type of system can be thought of as an electronic notepad or electronic blackboard, depending on the types of messages being sent. The system still qualifies as a facsimile system since a copy on paper of a written document is produced at a remote location. The only difference is that the original has not been prepared beforehand but is being produced at the time it is being sent through the system. One obvious use for this system is to send authorization signatures for financial transactions to a remote bank.

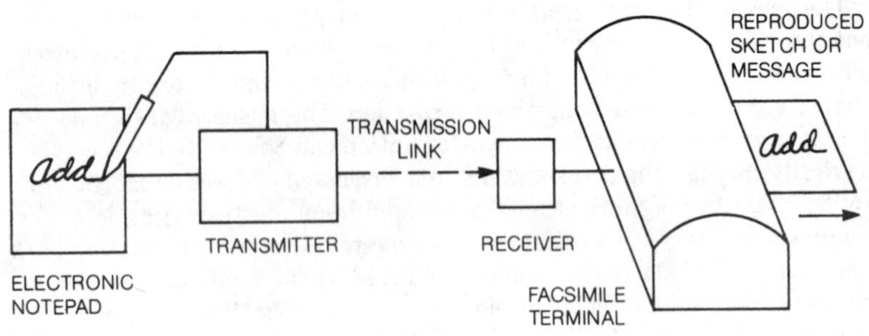

Figure 9-2. *Electronic Notepad System.*

A very close equivalent to the handwritten message terminal is the computer terminal that was discussed in Chapter 8. In these systems, illustrated in *Figure 9-3*, the computer or typewriter terminal at the sending station generates a typewritten message that is transmitted to a receiving station. At the receiver this message can be reproduced by a typewriter terminal or a printing terminal. This is not defined strictly as a facsimile system for two reasons. First, the input is restricted to typewritten information generated by a terminal operator or a computer. It is not possible to copy any arbitrary hand generated sketches or messages or to copy documents containing pictures or drawings. Second, the printing process does not provide a means to reproduce information in any other form than typed material. This system does share with a complete facsimile system the feature of providing a printed copy of the information on paper for filing or study over a period of time.

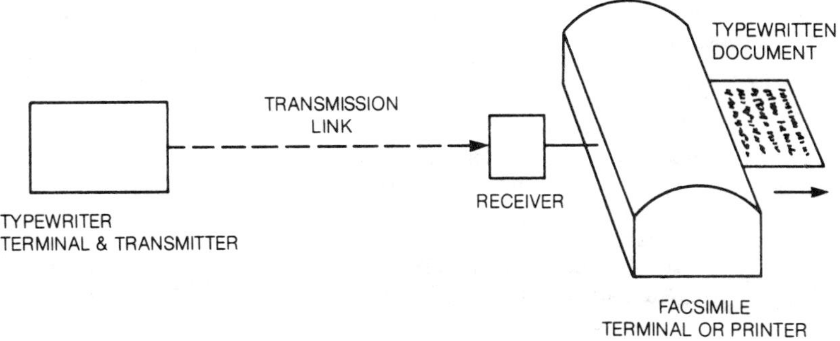

Figure 9-3. *Remote Typewriter System.*

The television systems of Chapter 7 also share some of the characteristics of the facsimile system. A simple block diagram of a system is illustrated again in *Figure 9-4*. In these cases visual patterns (pictures), which can be messages or drawings or combinations of these are converted to electrical signals and transmitted to a receiver. However, at the receiver, instead of providing copies of these patterns on paper or photographic film, the images are simply displayed on a television screen. Thus, since no hard copy is provided, strictly speaking this is not a facsimile system. Furthermore, most television systems are involved in the transmission of frames of visual patterns or pictures that successively follow each other in time and which may be changing in time. This does not occur in the transmission of the image of the single document in a facsimile system.

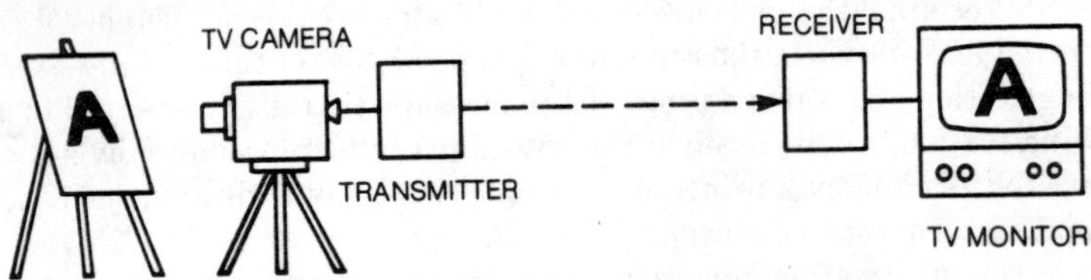

Figure 9-4. Television System for Transmitting Images and Messages.

FACSIMILE APPLICATIONS

Representative applications for which facsimile systems are ideally suited are listed in *Figure 9-5.* They are usually situations that require that a paper copy of a document or picture at one location be used at another location.

- Business Documents, Purchase Orders, Receipts, Invoices
- Engineering Sketches and Documentation
- Law Enforcement Information (Fingerprints, Records, etc.)
- Personal Messages and Pictures
- Library Materials
- Newsphotographs and Newspapers
- Weather Maps and Meteorological Information

Figure 9-5. Typical Applications of Facsimile.

Business

The most obvious example of the use of a facsimile system in the business field is when a receipt, purchase order, invoice, agreement, planning schedule or other business document is located in one part of the world and must be made available in another part of the world and there isn't enough time to wait for postal service or other type of delivery. A prime example occurs in newspaper publishing. News photographs which are taken in the afternoon in Paris must be made available in the United States in time to be published in the next morning edition of an American newspaper. Facsimile systems are ideally suited for this application. In some cases, the photographic plates for entire newspapers are sent by facsimile from a central publishing plant to publishing plants in various

regions of the country, thus allowing the material to be written and edited in an originating city, such as New York, but published and distributed throughout the country. The familiar wire services of Associated Press and United Press International have long been using facsimile systems to transmit news stories and news photographs to subscriber newspapers throughout the world.

Weather

Another very important and common use of facsimile is for transmitting weather maps to TV stations and airports across the country, and to ships at many locations at sea. This allows all the weather data to be collected and reduced at a central location and the resulting maps to be distributed to interested parties throughout the world.

Law Enforcement

Most readers are familiar with the use of facsimile by law enforcement agencies to transmit fingerprints, mug shots, driver's licenses, vehicle registration information and police records back and forth from local police organizations to state and national police agencies. The location of and true identity of suspected criminals can be verified very quickly and efficiently using such techniques.

Libraries

Another ideal use of facsimile is in the distribution of library documents and archives from a central location to locations throughout the world. In this way, a central library can make available to the public at regional and remote locations copies of expensive and rare original books and documents that cannot be moved.

Personal

Finally, personal documents needed, such as birth certificates, copies of checks, and copies of wills can be sent by facsimile when an individual must get documentary evidence to a distant location in a minimum of time. In most of the applications listed in *Figure 9-5* the concept of long distance transmission of important documents has been emphasized. As energy becomes increasingly scarce, personal delivery of documents to an individual by the mail service or other special delivery service could become unfeasible, leaving the only possible way to transfer these documents from one location to another to the electronic means offered by facsimile.

Since any document or picture can be sent by facsimile, it might be assumed that facsimile involves complex technical procedures or principles that are difficult to understand. This is not true. Most of the basic concepts and techniques involved in facsimile systems are very simple to understand.

HOW DOES FACSIMILE WORK?

The basic components of a facsimile system are shown in *Figure 9-1.* There must be a device that converts the visual (optical) pattern of the source document into the electrical signals that represent that pattern. The electrical signals are transmitted using the techniques of Chapters 2, 3, and 4. At the receiver these electrical signals are converted back into the corresponding visual (optical) pattern of the source document. The new techniques of the facsimile system that must be understood are those used to convert the optical patterns from a two-dimensional source document into electrical signals for transmission, and then the inverse conversion of these signals back into a facsimile of the original input after they are received. These techniques will be examined beginning with the conversion that occurs in the transmitter, which can be called encoding the optical information.

Encoding Optical Information to Electrical Form

As the source document moves through the facsimile sending unit, patterns of light varying in level to represent the lightness or darkness of the picture or document being examined must be converted to electrical signals. The varying light levels can vary the amplitude or frequency of an electrical parameter such as voltage or current in order to accomplish the conversion. Fortunately, devices exist that will perform this conversion easily.

Photodiode

An example of one such device, the photodiode, is shown in *Figure 9-6.* When light strikes the photosensitive portion of the diode a current is produced through the diode. The more intense the light is, the larger the diode current. As shown in *Figure 9-5,* connecting a circuit so that current can flow through a resistance to produce a voltage change at the input to an amplifier detects the light intensity changes. The output of the amplifier can then be a voltage which is proportional to the light intensity hitting the diode. By using electronic amplification, this voltage level can be increased enough to modulate a transmitter in one of the ways described in Chapter 3 and 4 so that the light intensity information can be sent to a remote receiver.

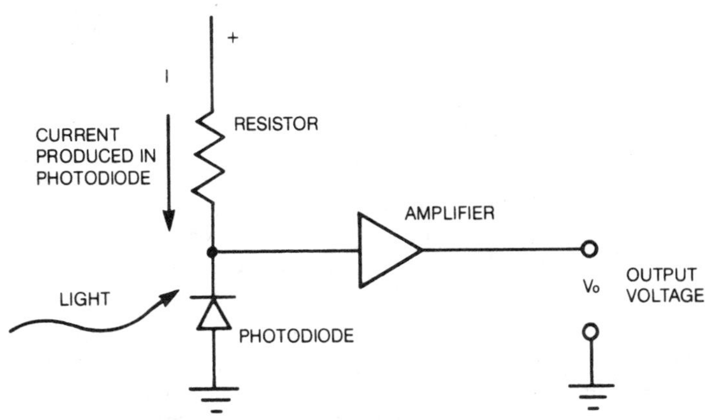

Figure 9-6. *Photodiode Light Detection.*

Photomultiplier Tube

Another device that is used to perform the conversion of light intensity to electrical signals, in this case current, is the photomultiplier vacuum tube. This tube, shown in *Figure 9-7*, has an amplifying mechanism built right into the tube. As light hits the light sensitive photoelectrode within the tube, electrons are emitted from the surface of this electrode. Once emitted, the electrons are attracted by a high voltage toward the next electrode of the tube. In so doing, they are also accelerated. When these fast electrons hit the next electrode, they cause more electrons to be emitted from the surface than the number that arrived at the surface. The emitted electrons are attracted and accelerated to the next electrode with a high voltage, as before. At each electrode, the number of electrons emitted and sent to the next electrode are more than the ones hitting the emitting electrode. By the time the electrons reach the last collecting electrode, there are millions of times more electrons than the number emitted from the photoelectrode. As a result, as shown in *Figure 9-7*, the current that flows through the resistor connected to the collecting electrode can be relatively large, even for very low light intensities. The photomultiplier vacuum is thus a very sensitive light detector. Very little additional electronic amplification is required to provide the voltages needed by the transmitter circuits. While most of the needed amplification is provided by the photomultiplier tube, it does have the disadvantage of requiring high-voltage power supplies in order to operate properly. High-voltage power supplies are more costly, therefore, the equipment is more expensive. Also, extra protection is required so that operators and repairmen are not harmed by the high voltage.

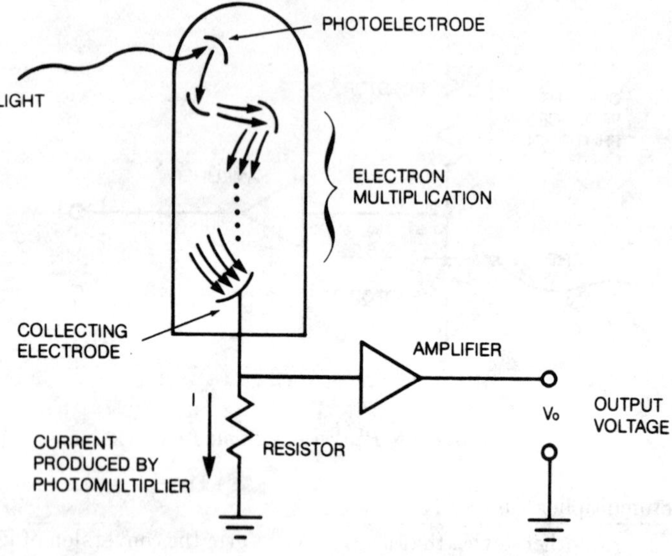

Figure 9-7. *Photomultiplier Tube Light Detection.*

Spotlighting

Regardless of which photodetector is used to convert light intensity to electrical current and voltage, there must be some means of restricting the light intensity variations to selected portions of the source document being transmitted. Either the light source must be focused onto selected portions of the input document, or the photodetector must be focused to look at only the selected portion. In other words, either the light illuminating the document must illuminate only the portion of the document that is to be converted to electrical form, or (if the document is uniformly illuminated) the photodetector must be focused only on the portion of the document that is to be converted to electrical energy.

The first approach of restricting the illuminated area is a "spotlight" approach and is illustrated in *Figure 9-8.* The spotlight can be formed by focusing a beam of light with a lens onto the paper being examined. In this way only a small portion of the document is being illuminated. The average light level of the illuminated spot causes a reflected light of a certain intensity. By detecting this reflected light with a photodetector, the average lightness or darkness of the portion of the document being illuminated is converted to a voltage at the output of the photodetector circuit. If the spot of light is on pure white, the maximum reflected light will occur, causing the maximum electrical voltage to be generated. If the area illuminated is black, as would be the case if it illuminated a period on this page, the reflected light would be minimal, and the corresponsing electrical voltage out of the photodetector circuit would be low. A gray area in a picture would cause a voltage in between these two extremes. Thus, with this technique, one small portion of the picture, typically smaller than a period on this page, is converted to its equivalent electrical voltage.

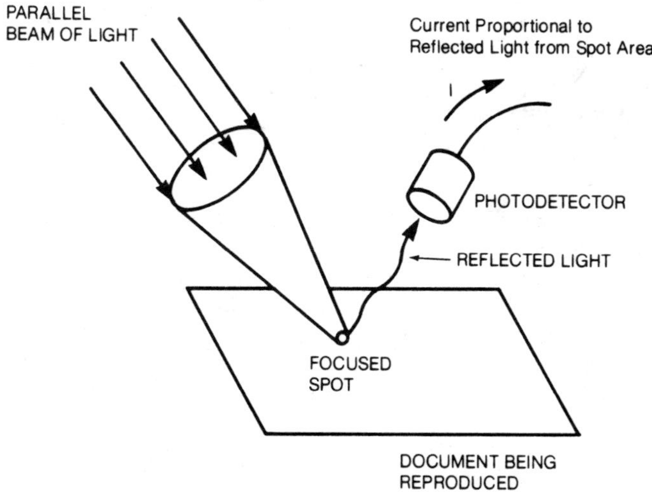

Figure 9-8. *Spotlighting the Picture Detail.*

Focusing the Photodetector — Aperture Approach

When the photodetector is focused, the approach is to uniformly illuminate the entire source document being examined, while restricting the "view" of the photodetector to a small area – an area as before, about the size of a period on a printed page. This approach, called an aperture approach, is illustrated in *Figure 9-9*. With the page uniformly illuminated, light is being reflected from the full surface area. If a lens-aperture-detector configuration of the type shown in *Figure 9-9* is used, the detector will receive light only from a small area on the paper. Any light coming from outside this area will be blocked by the aperture plate and will not be detected by the photodetector.

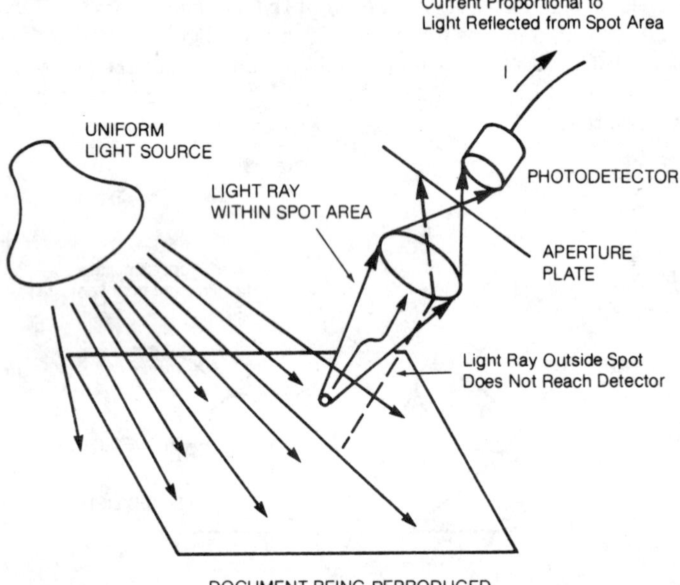

Figure 9-9. *Aperture Method of Focusing on the Picture Detail.*

Either the spotlight or aperture method can be used, but the spotlight method requires less light power than the aperture method for a given sensitivity. Also, the spotlight method can be achieved by using a laser beam which offers very small spot size and good control over the light intensity of the source. In either the spotlight or aperture method, the intensity of the source light must be held constant so that the detector voltage is indicating the relative light level variations from the documents in the area examined and not indicating the variations in the source of light. The current variations are amplified to the required levels for transmitting in the manner shown in *Figures 9-6* and *9-7*.

Spot Scanning Concepts

Either of the techniques of *Figures 9-8* and *9-9* are useful for converting the light level reflected from a small area of a source document. However, all portions of the document or picture must be examined and converted. This means that the area that is examined must be varied so that the entire surface area can be converted. The most common way of causing each area of the paper to be examined successively is to scan the spot across the surface of the paper in a regular pattern. This is called a scan pattern or raster. A typical line-by-line scan pattern is shown in *Figure 9-10.*

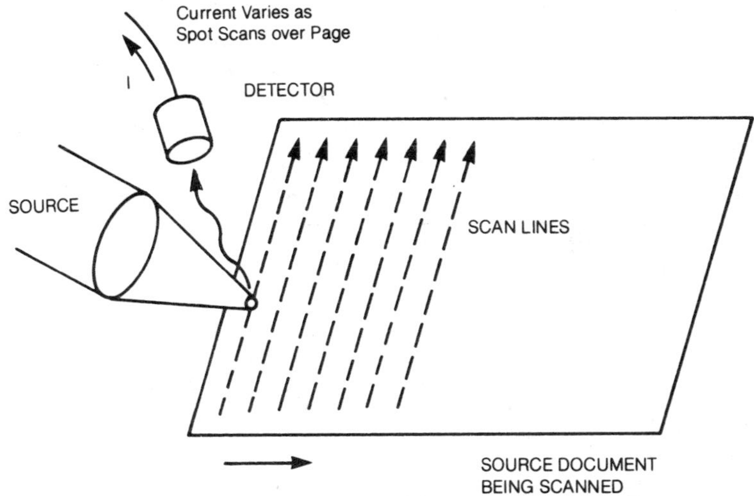

Figure 9-10. *Light Spot Scanning of a Document.*

Spotlight and Aperture

In this case the spotlight approach is shown. The spot is started at one edge of the paper in the corner of the paper, moved across the page to the other edge of the page, moved down one spot diameter as it returns to the starting edge of the paper, and then moved across the page again on the new line. This scanning is repeated until the entire surface area of the page has been illuminated by spot on a line-by-line basis. A similar scan is used when an aperture detector is used, by moving the focused spot of the photodetector over the page in the same line-by-line pattern.

Scanning Details

This scanning approach is much the same as that used in the television systems discussed in Chapter 7, except that the scanning tends to be mechanical instead of electronic, so that the facsimile scanning rate is much slower than the television scanning rates. In either case the overall picture is being broken down into thousands of individual black-to-white light level readings, each taken over a small spot size. This is illustrated in *Figure 9-11. Figure 9-11a* is for circular spots; *Figure 9-11b* is for rectangular spots. The photodetector is providing a voltage output that is proportional to the average light level present in the picture at the spot being detected. If circular spots are used and the area of one scan line is tangent to a second line as shown in *Figure 9-11a,* some portions of the paper are never examined by the photodetector. This is not a problem with rectangular or square spots as shown in *Figure 9-11b.* The rectangular spot can be formed either by using a rectangular hole for the aperture method or by using a rectangular source beam of light with the spotlight method.

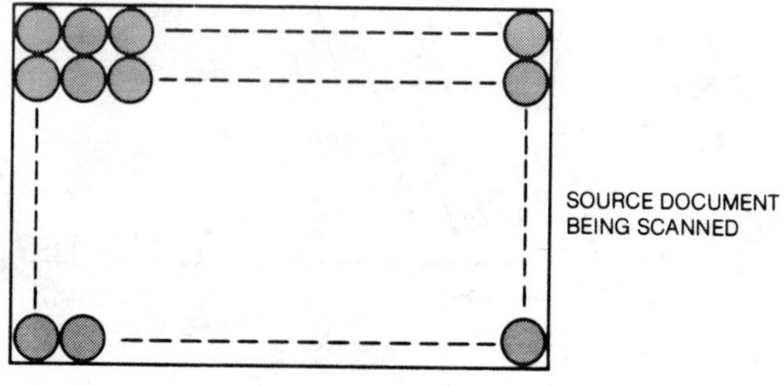

a. Circular Spot Picture Elements.

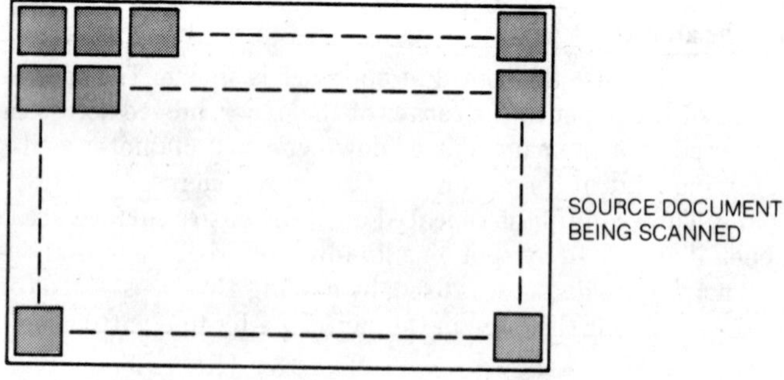

b. Rectangular or Square Spot Picture Elements.

Figure 9-11. *Dividing Source Document into Small Picture Elements for Scanning.*

Scanning A Character

Regardless of whether a circular or rectangular spot is used, the effect of scanning a line of paper containing alphabetical typewritten characters can be seen by examining *Figure 9-12.* In this case the spot is scanning across the capital letter H. Until the spot reaches H the white of the paper is providing a maximum reflected light and the voltage out of the photodetector is at a maximum level. If the scanner is on line n in *Figure 9-12a,* the electrical voltage will drop to a low level when the spot is centered over either leg of the H, since this is a dark part of the page. Of course, if the spot is larger (wider) than the width of the legs, this voltage will not be as low as it would be if the spot width was less than the width of the H leg. In the latter case, the photodetector would see almost no reflected light and would provide a minimum output voltage level. When the spot width is larger than the leg width, some of the white paper is reflecting even when the spot is centered over the leg, so the voltage output of the detector will not be as low as it is for the spot reflecting from a total black area.

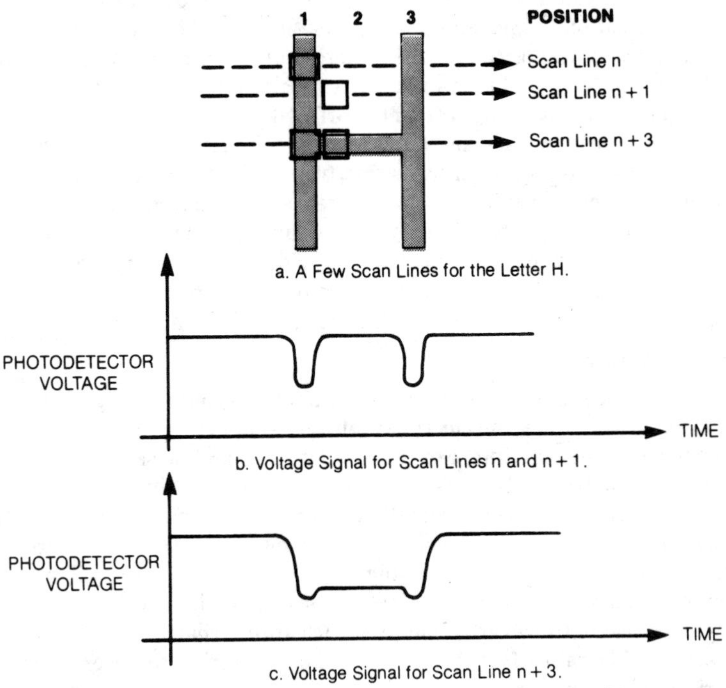

***Figure 9-12.** The Relationship Between Optical Image and Electrical Signals in Facsimile Scanning of Characters.*

In either case, the voltage output of the detector will drop momentarily as the spot crosses the two legs of the H as shown in *Figure 9-12b.* This same voltage pattern versus time would occur for both the n and n + 1 scan lines shown in *Figure 9-12a.* The output is different for scan line n + 3. The spot would see mostly black all across the center of the H, with the only white coming from areas in which the rectangular spot extends above or below the black horizontal bar of the H that connects the two vertical black bars on each side. In fact, the spot would be seeing the most black at positions 1 and 3 in the scan, and a little less at position 2 in the middle of the H. Thus, the detector output voltage for line n + 3 would look like *Figure 9-12c,* with slight dips in the signal when the scan spot is at positions 1 and 3.

In both *Figures 9-12b* and *9-12c,* the effect of a spot moving across the character at a uniform rate results in a gradual slope in the voltage change from one level to another. This is due to the way the spot sweeps across a dark area from a light area. *Figure 9-13* shows the result by illustrating such a light-area-to-dark-area transition and the resultant detector output voltage plotted against time. Before the spot enters the dark area (Position a) the voltage is at a maximum. When the spot is over the total black area (Position c) the voltage is at a minimum. When the spot is half over light and half over dark areas (Position b) the voltage is at a midway point. For the character on the printed page there is an abrupt change from white to black and vice versa. For the electrical waveform of *Figure 9-13* there is a gradual transition from white to black so the reproduction is not exact but still very acceptable for a large majority of all types of source documents. The only way to cause the electrical voltage to make a more abrupt change is to make the width of the spot narrower. In general, the smaller the spot is made, the more accurately the electrical signals from the detector represent the visual pattern on the paper.

Scanning Pictures

If the document being scanned is a picture containing gradual variations in gray tone, the electrical voltage from the detector should vary similarly along the scan line. For example, if the scan line sees a gradual variation from black on the starting side of the page to white at the end of the scan, the voltage should gradually rise from a low level at the start of the signal to a high level at the end. This is shown in *Figure 9-14.* Since the optical information is gradually changing, the detector voltage will be able to generate a similarly changing electrical signal, with little dependence on spot size or shape. However, in pictures with abrupt changes in tone, the spot size would again affect the accuracy of the electrical representation of the original information.

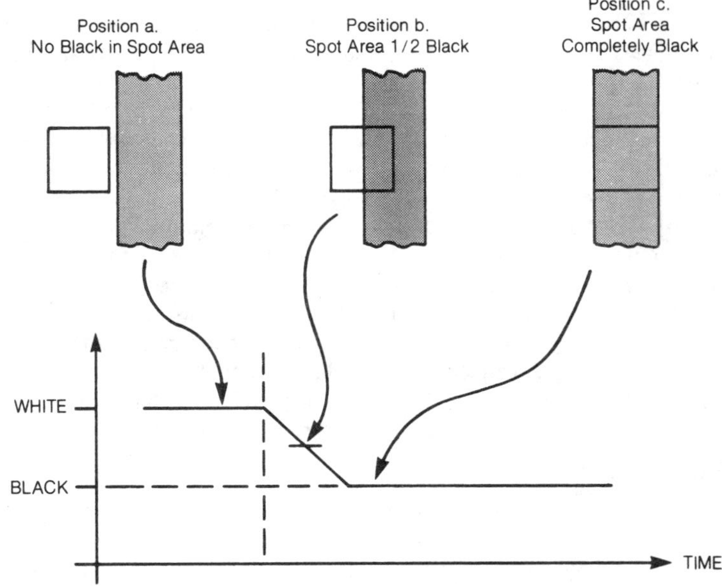

Figure 9-13. *Electrical Signal Generated by a Scanning Spot Crossing a Black Bar.*

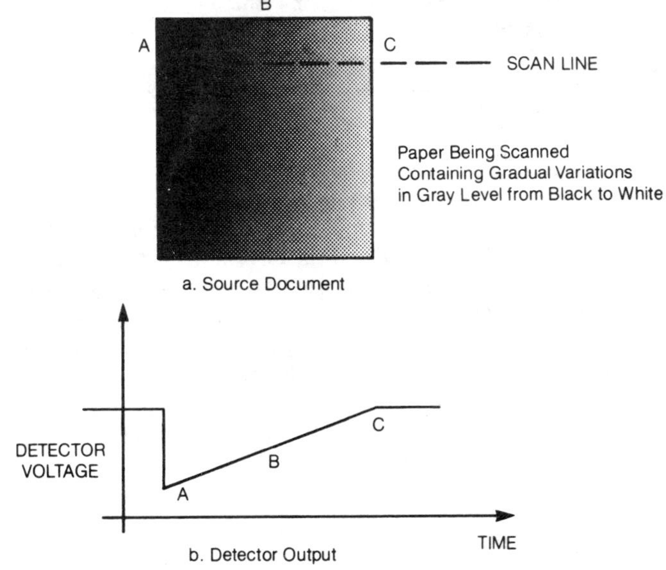

Figure 9-14. *Electrical Signal for A Scan Line with Gradual Variations in Gray Level.*

If the spot size is reduced to increase the accuracy with which the voltage variations represent the picture variations, the picture will be transmitted with sharper detail, but each scan line will examine a smaller portion of the paper, and more scans will be required to cover the entire length of the page. For example, in *Figure 9-11b,* if the length of the spot is reduced to 0.01 inch (one-hundredth of an inch), 100 such scan lines will have to be made to cover one inch of the length of the paper. For an 8½″ by 11″ sheet of paper, 1100 such scan lines will have to be made. This has compounding effects. Generally, the sharper the picture detail required by the system, the more scan lines per inch (LPI) required, the more costly the scanning equipment, and the more precise and expensive the optical equipment. Further, the cost of transmitting the information will increase. Before discussing the affect of the resolution requirements (which will be done later in the chapter), let's examine some of the techniques of performing the scanning of the document to be transmitted.

Scanning Techniques

In order to cause the spot examined by the photodetector to move across the paper in some orderly manner, either the paper must be moved relative to a fixed position photodetector, or the focused photodetector must be moved relative to a fixed piece of paper. Both techniques are used in facsimile systems. In addition, whatever is moved may be moved either continuously or digitally (in discrete steps). Both of these approaches also find application in facsimile systems. Some systems use a combination of these alternatives.

Paper Mounted On Cylinders

One of the common early methods of providing a scan of the paper surface is illustrated in *Figure 9-15.* In this scheme, the paper containing the picture or information to be transmitted is wrapped around and fastened to a cylinder. Then a motor drives the cylinder at a certain speed in revolutions per minute (rpm). The photodetector-light source assembly provides a spot light source and the photodetector to detect the light level reflected from the traveling spot on the paper as the cylinder rotates. The detector assembly moves down the paper as the cylinder rotates. The detector assembly moves down the paper at a linear rate so that at the end of one rotation of the cylinder the line scan position has moved down one spot height. When the photodetector scans the last line, the machine stops, the source document is removed and another one mounted on the cylinder for the next transmission.

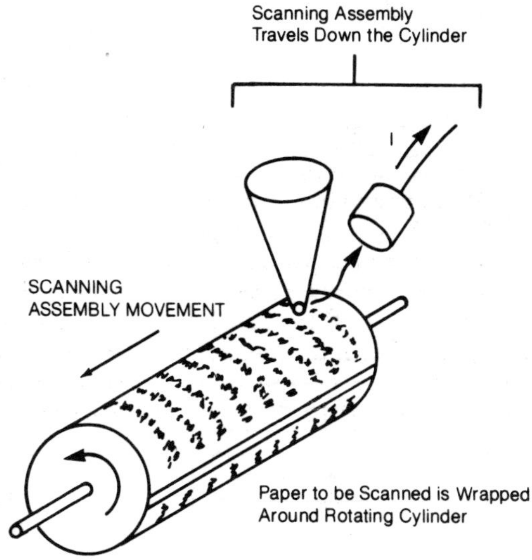

Figure 9-15. *Cylindrical Scanning System.*

The type of scan pattern produced by this method is shown in *Figure 9-16*. Since the paper is wrapped around a cylinder, the next scan line begins where the last scan line ended. Thus, the system avoids having to provide a "flyback" (return of the spot from end of page to beginning of page) of the scanning assembly to the beginning of the next line, as would be required in flat paper scanning techniques or as required in the television system scanning techniques. Further, by encoding the shaft or the edge of the cylinder with a light reflective band, the machine can generate end-of-line/beginning-of-line information that can be used to sychronize the transmitter and receiver scanning processes. Such synchronization can be very important in the quality of the reproduction at the receiver terminal.

There are many variations of the cylindrical system of *Figure 9-15*. For example, the light source-detector assembly could be held in a fixed position and the cylinder moved along the axis (longitudinally) under the assembly as it rotates. The system shown in *Figure 9-15* requires the document to be wrapped around the cylinder and fastened when it is loaded. A document on stiff paper or very flimsy paper could be damaged by this approach. Even with strong thin paper, the loading is not as convenient as in other types of scanning systems.

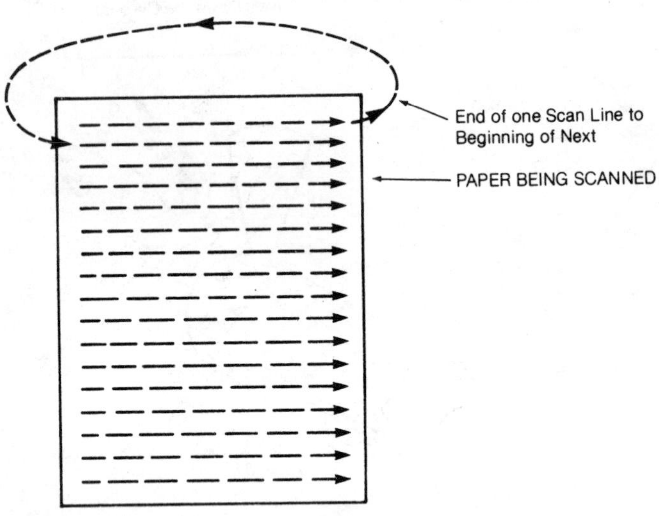

***Figure 9-16.** Typical Scan Pattern or Raster.*

Flat Paper Scanning

One type of non-cylindrical scanning structure is illustrated in *Figure 9-17.* In this system the paper is fed through a roller system while the sheet lies flat on a scanning table. At the same time, the source/detector assembly is moved across the sheet to scan one line. It must be returned back to the starting position of the scan and the process repeated until all lines have been scanned. All of this time the paper is traveling through the roller assembly. The scan pattern is similar to that shown in *Figure 9-16,* though the start of the next line may be offset slightly from the end of the last line due to the finite flyback time. This flyback time could be the limiting factor on how fast the paper may move linearly through the system. The advantage of the system is the ease in which documents can be loaded into the scanner. This type of system also allows the use of many more different types of paper than does the cylindrical system.

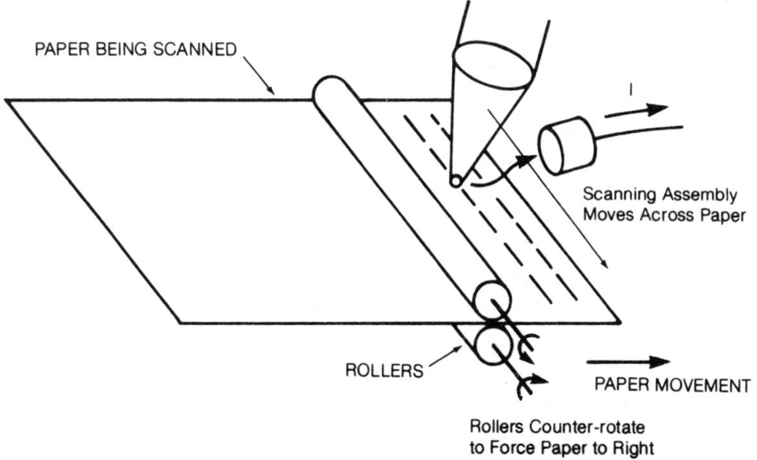

Figure 9-17. *Non-Cylindrical Scanning Mechanism.*

Slit System Of Spot Scanning

One way to reduce or even eliminate the flyback time of the scan line in the system of *Figure 9-17* is to use a rotating wheel to generate the flying spot. One approach is shown in *Figure 9-18*. Here a moving vertical slit and a stationary horizontal slit form an aperture that light passes through from a source. Since the vertical slit is moving, the light spot moves across the page and can be followed by the photodetector. By placing a new vertical slit to start on the left when the last vertical slit leaves the right side of the horizontal slit, the flyback time can be reduced to zero. This is accomplished using a band with vertical slits spaced apart at a distance equal to the width of the paper and having the light source inside this band. The same scan pattern is generated as for the cylindrical system. Some systems realize the same effect by having a spiral slit cut in a rotating wheel. This causes the vertical slit to move past the horizontal slit at the desired rate. This approach also reduces flyback time to negligible levels.

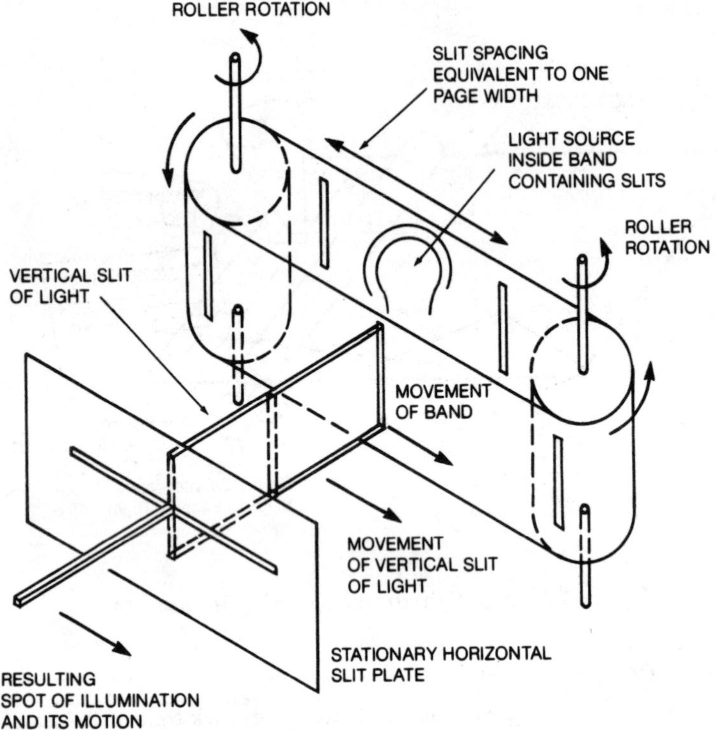

Figure 9-18. *Linear Scanning Technique with No Flyback Time.*

Stationary Source Documents

In both of the systems described thus far, mechanical scanning of the optics along one dimension is used along with mechanical motion of the document along the other dimension to provide the scan of the complete surface in a pattern like that of *Figure 9-16.* If it is desirable for the document to remain stationary, either for ease in loading or some problem with the type of visual source information, the scanning spot must be moved in both directions. One way of doing this is to provide a system of rotating mirrors as shown in *Figure 9-19*. A narrow beam of light is input to the mirror system. This beam could be produced by a laser or by conventional sources. The first mirror rotates about one axis, causing the beam to scan along one axis on the paper. The other mirror rotates at a slower rate around the other axis, causing the fast scanning beam to move down the page to produce an overall raster of the form of *Figure 9-16*. Again, some technique must be used to prevent excessive flyback time.

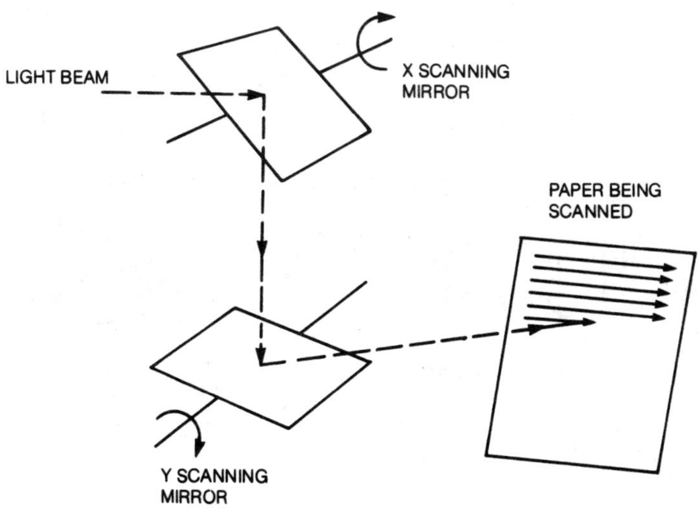

Figure 9-19. *Mirror and Light Beam Scanning System.*

Multifacted Mirror Scan

One possible solution to this problem is to use a multifaceted mirror structure of the type shown in *Figure 9-20a*. At the initial point in the rotation (solid lines) the beam is reflected to Position 1 (*Figure 9-20b*). As the mirror rotates, the reflection would move to Position 2. When the next mirror face appears a very short time later the reflection returns to Position 1. Thus, the beam will scan from Position 1 across the page to Position 2 and then start back at Position 1 for the next line. In the meantime, a second bevelled mirror can be causing the scan down the page in a similar manner. The second mirror only causes one complete down-the-page scan so that the features of *Figure 9-20* are not necessary for this mirror. The speeds of the two mirrors must be synchronized closely, both with each other and with a similar scanning mechanism at the receiver.

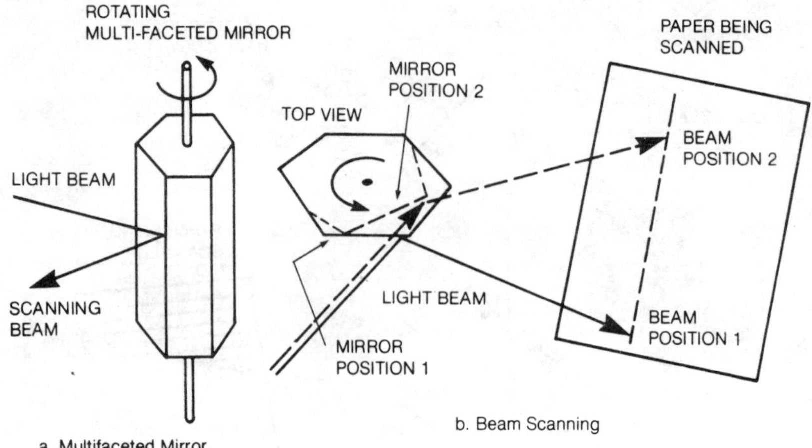

Figure 9-20. *Non-Flyback Mirror Scanning Technique.*

Electronic Scanning

A simpler and mostly electronic scanning approach is shown in *Figure 9-21.* Here a high intensity CRT is used to provide a scanning spot to traverse an x-y raster. The system has the advantage of no mechanical moving parts, offering the option of very high scanning speed. It is very useful for directly exposing photographic film at the receiver which will be discussed later when reproduction techniques are covered. The CRT system also has a finite flyback time, but because the scanning is done electronically, it is much faster than any mechanical system and not an important limitation of this system.

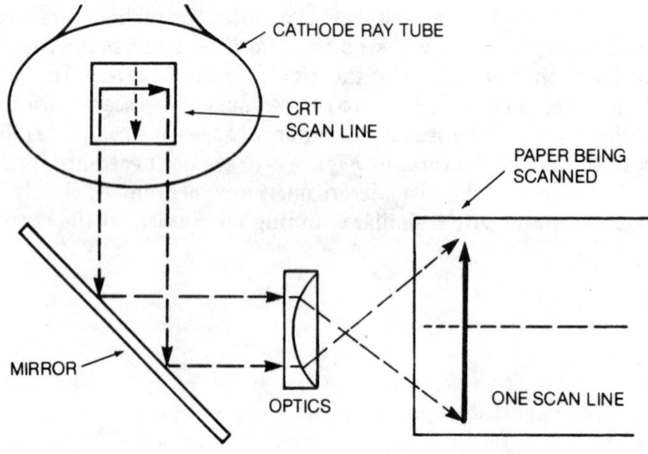

Figure 9-21. *CRT Scanning System.*

Composite Detector Output

Figure 9-22 is a composite output signal from the photodetector for a complete scan line. Each scanning system discussed provides an electrical output signal that provides a beginning of the page synchronization pulse and a start of a line synchronization pulse. These are shown in *Figure 9-22a*. They are at the white or highest light intensity level. After these occur, the scan-line signal level corresponds proportionately to the visual pattern that is on the source document.

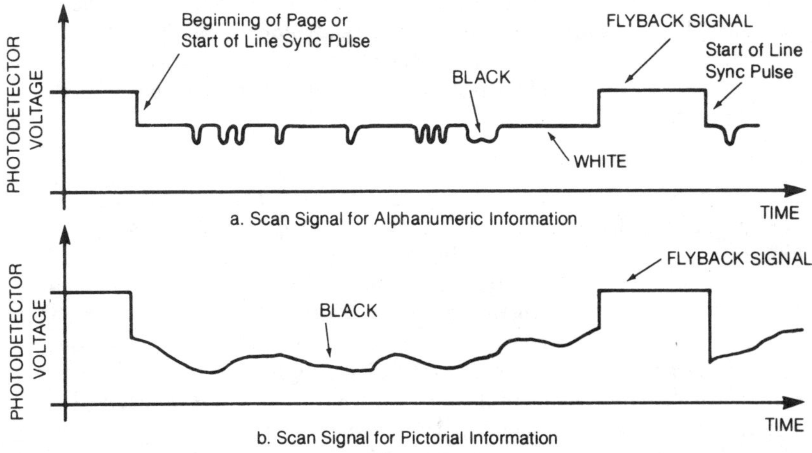

***Figure 9-22.** Electrical Signals Resulting from Typical Documents.*

If the document is printed material, this signal pattern will look like that shown in *Figure 9-22a* and the pattern of light and dark areas follow the form discussed in *Figure 9-12*. If a picture is being scanned with continual variations in gray (*Figure 9-14*), the electrical signal similarly will be a continuously varying voltage of the form shown in *Figure 9-22b*, with the higher voltages representing white areas of the picture and the lower voltages representing the dark areas of the picture (or vice versa, depending on the electronic circuitry used). These signals must be modulated onto a carrier and transmitted to the receiver by the techniques of Chapters 3 and 4. The parameters that affect this transmission and the type of modulation used will be discussed shortly. At the receiver the transmitted signal is demodulated to recover the waveforms similar to those in *Figure 9-22*. To reproduce the copy of the source document, the electrical voltage patterns must be reconverted back into visual form as permanent marks on a piece of paper or on a photographic film. These reproduction techniques will be considered next.

FACSIMILE REPRODUCTION TECHNIQUES

The techniques that are available for converting electrical signal information into permanent marks on a piece of paper differ in their convenience and cost, and in their ability to generate different levels of gray tone from white to black. However, all share the common requirement for a scanning mechanism at the receiver that can be synchronized to the scanner at the transmitter terminal. This is necessary so that the spot marked at the receiver on the reproduced document corresponds to the point that was sensed by the photodetector in the sending unit. For this reason, the same scanning techniques used at the sending unit are used at the receiver. These scanning techniques are already understood; therefore, to complete the system, the techniques used to convert the electrical signals received back to the visual patterns of information required for the copy must be understood.

Electrostatic Reproduction

A cylindrical scanning reproduction system that is similar mechanically to that of *Figure 9-15* is shown in *Figure 9-23*. This figure also illustrates one method of converting electrical information into different levels of gray on a piece of paper. The reproduction technique illustrated in *Figure 9-23* is called electrostatic printing. It is the type used in many office copiers. The main component of this system is a cylindrical drum whose surface charge is controlled by the intensity of light that falls on it. Drums made with a surface of selenium have this characteristic. The received signal pattern is used to modulate the intensity of a light beam and scanned over the surface of the drum with the same spot size as the sending unit. This places a pattern of charge proportional to the signal on the drum surface. Once a varying light pattern has been converted to a corresponding varying charge pattern on the drum surface, the drum surface is passed by a powder that is attracted to the drum surface in proportion to the charge on that surface. After the drum has the powder on the surface it rotates to transfer the powder to a piece of white paper that is being fed by rollers past the drum. As the paper exits the system, heat and pressure are applied to the powdered paper to fix the reproduced image onto the paper. Where the light is more intense (white areas) the charge and powder accumulation is heavy, and the corresponding portion of the final reproduction is dark. If the electrical signal being received from the transmitter has the forms of *Figure 9-22* in which high voltages indicate white areas and low voltages indicate dark areas, then high voltages must be converted to a dim light spot hitting the drum surface and low voltages must cause a bright light spot on the drum surface in order to reproduce the source document. This can be done relatively easily once the source of light is decided on.

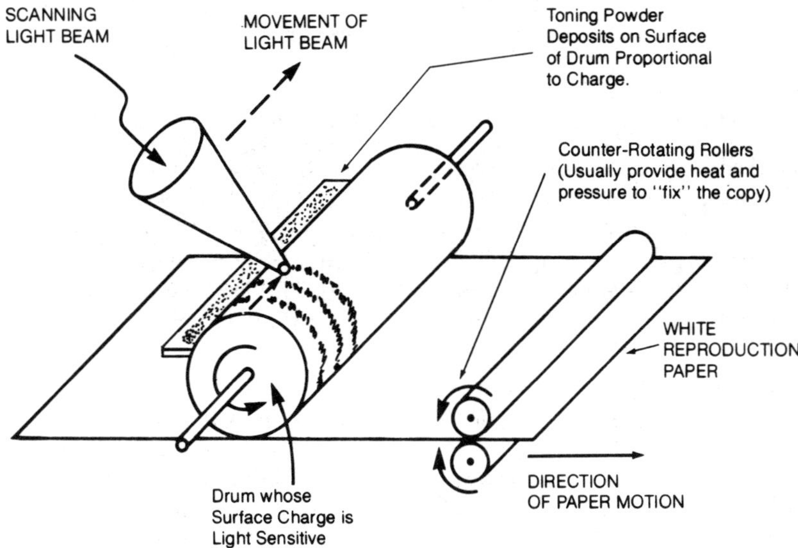

Figure 9-23. *Electrostatic Reproduction Technique.*

One light source that has built-in scanning capability is the CRT or cathode-ray tube. By having a CRT generate the light pattern that scans across the drum, it is simply a matter of converting the input signal to CRT spot intensity as the CRT beam sweeps across the face of the CRT in the time representing one scan line. Adjusting the drum speed to match the drum or line scanning rate that was used at the transmitter synchronizes the system. This system can be very expensive, requiring special materials and close mechanical design tolerances. Because of the powder, filling the machine can be very messy.

The advantages of the system are these: either high- or low-quality paper can be used in the final reproduction. The system has good copy definition. Very gradual variations in the source light and dark areas can be reproduced fairly accurately. This means pictures can be transmitted and reproduced with very good quality. Typewritten documents can be reproduced that are almost as good as the original. In summary, the electrostatic system provides a system that offers the best accuracy in gray level reproduction and produces a very high-quality copy when good-quality paper is used. It is relatively easy to operate but is likely to be more expensive.

Electrode Type Reproduction

Simple systems that offer a wide range of optical gray levels and fairly high scan rates are the systems that use the electrode contact elements of *Figure 9-24* for reproduction. *Figure 9-24a* shows an electroresistive paper that turns dark in proportion to the current that is passed through it (high current = dark; low current = light). By controlling the amount of current so that it reproduces the signals of the form of *Figure 9-22b*, the resulting document can provide the same detail and gray scale features of the original picture. Either the cylindrical or rectangular scanners of *Figures 9-15* or *9-17* can be used to move the electrode over the specially treated paper in the appropriate raster pattern. This system has two main disadvantages: 1. Special reproduction papers must be used. 2. Final copies may be slightly colored.

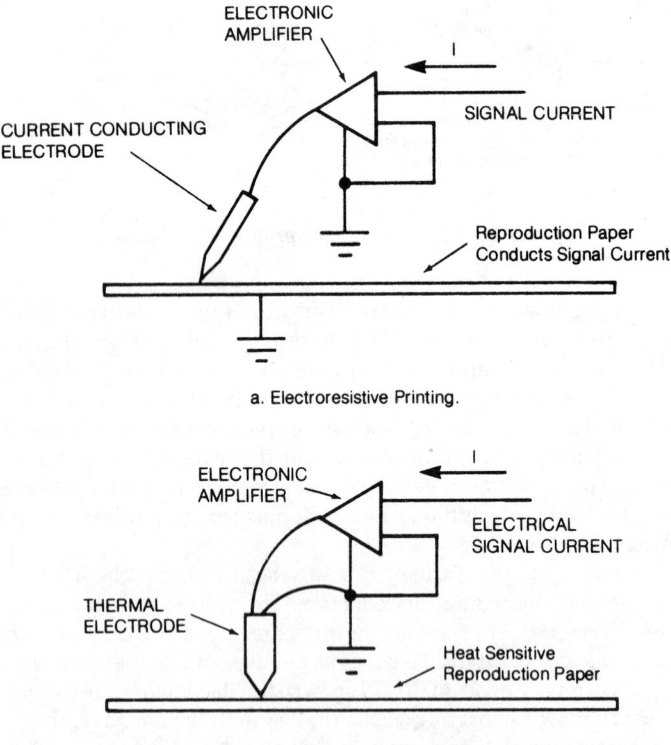

Figure 9-24. *Electrode Type Reproduction Techniques.*

The same basic principle is used for the thermal print head of *Figure 9-24a;* however, in this case, the specially treated paper darkens when heated (high heat = dark; low heat = light). The temperature of the point of the head must be varied in accordance with the *Figure 9-22* electrical waveforms received. Picture details can be reproduced fairly well by this process using relatively inexpensive paper, again using a cylindrical or rectangular scanning system.

Ink-jet Printing

The ink-jet printing system of *Figure 9-25*, developed for high speed, high contrast printing of documents in computer systems, is also a useful facsimile reproduction. The amount of ink that is sprayed through the nozzle onto the paper is controlled by the electrical signal and the pressure transducer. Thus, if a signal of the form of *Figure 9-22a* is sent to the transducer, ink spots will be sprayed for every dark level pulse in the line scan signal. No ink is sprayed when the signal is at the white level. This process can be used on a wide range of papers and is primarily useful for reproduction of printed material and drawings that have high contrast material on the source document rather than gradual gray tone areas.

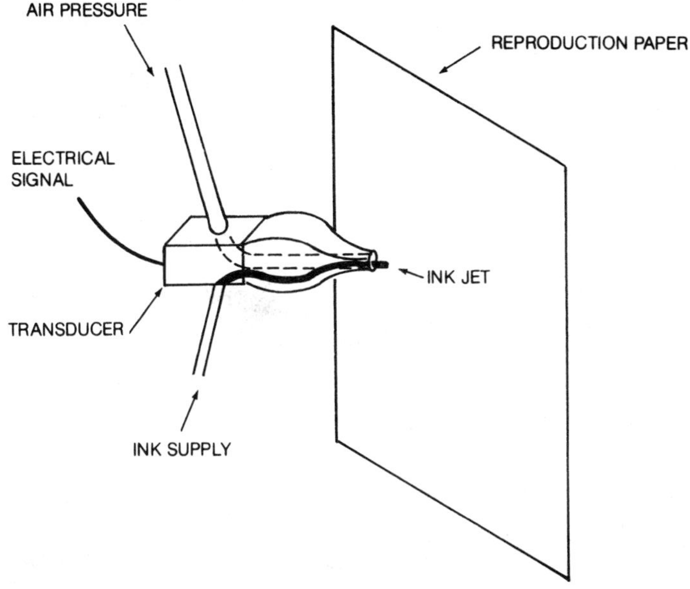

Figure 9-25. *Basic Ink-Jet Printing Technique.*

Photographic Copies

All of the reproductive techniques discussed thus far are means used to form an image on paper to provide a permanent copy of the original document. Photographic copies also can be provided at the receiver. If microfilm or offset printing masters are sent from a central location to a remote location, such photographic techniques would be required. For this technique a narrow beam of light whose intensity can be varied is needed. This beam must be scanned across the photographic film in the way the original was scanned at the transmitter. One possible approach is to use a laser with a light modulator and a mirror scanning system of the form illustrated in *Figure 9-19* and *9-20*. The modulator must be capable of varying the light intensity in accordance with the line-scan signal (*Figure 9-22*).

A simpler system is shown in *Figure 9-26*. It uses a CRT tube to provide both the scanning and beam intensity modulation. The CRT beam is scanned across the face of the CRT with the same scan rates and patterns that were used at the transmitter. The beam intensity is varied using the scan-line signal. The image thus produced on the face of the CRT is focused on the microfilm, polaroid film, or photographic plate, in order to expose the film with the image. Processing and printing the film reproduces a copy of the original source document at the receiver. Depending on the quality and detail required, this can be an involved and expensive system requiring special equipment and further processing, so it is not used in situations where reproductions on standard or special paper are required.

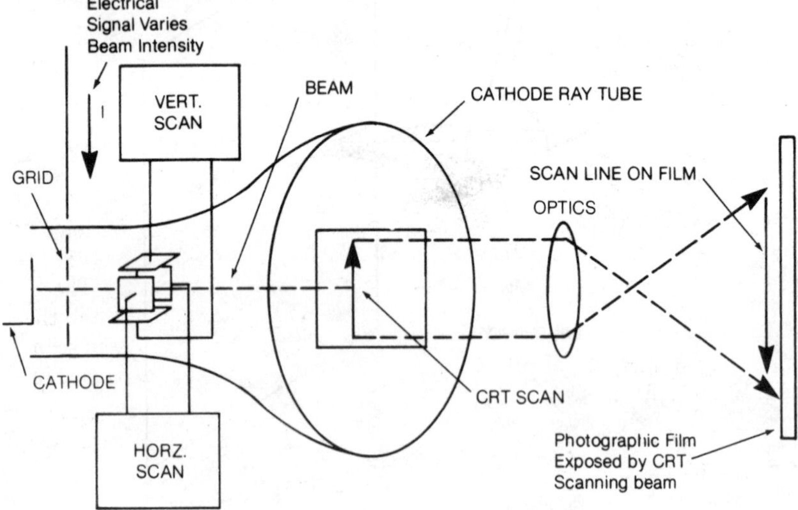

Figure 9-26. *Photographic Reproduction System.*

Which reproduction system is used in a given facsimile system depends in part on the type of document being reproduced. Some systems are suitable primarily for reproduction of printed material, while others have more capability and can reproduce detailed pictures as well. The transmission link used between transmitter and receiver depends on system parameters such as speed and cost. In order to be able to further evaluate systems the characteristics of typical systems will be discussed.

WHAT ARE THE CHARACTERISTICS OF TYPICAL FACSIMILE SYSTEMS?

Facsimile systems can be characterized by certain basic parameters which in turn affect the requirements of the communication system. The communications options that are most commonly used in facsimile systems are listed in *Figure 9-27*. The basic modulation systems used are amplitude modulation, frequency modulation, and pulse code modulation. The transmission is commonly over telephone lines, though radio transmission is necessary in many circumstances, such as sending weather maps to ships at sea. When radio transmission is used, the bandwidth requirements of the facsimile information are not too important. When telephone lines are used, the bandwidth requirements become an overriding consideration, with direct effects on transmission costs.

Modulation Technique	Transmission System
Amplitude Modulation	Cable Transmission
Single-Side-Band AM	Radio Transmission
Frequency Modulation — FSK	
Digital Coding — PCM	

Figure 9-27. *Communications Systems Alternatives for Facsimile.*

Basic System Parameters

Many of the facsimile system features and operating conditions directly affect the information bandwidth. The basic system parameters used in describing facsimile systems are summarized in *Figure 9-28*. The parameter LPI affects both the bandwidth needed for transmission and the resolution required to produce a good copy. It is the number of system scan lines used per inch as shown in *Figure 9-28*. The more lines per inch used in the scan, the higher the resolution of the picture.

Resolution in Lines per Inch (LPI)
Scan Rate in Lines per Second (LR)
Length of Scan Line (L)
Length of Paper (LS)

Figure 9-28. *Basic Facsimile System Parameters.*

Two other parameters that must be considered when computing system bandwidth (BW) requirements are the length of the scan line (L) and the scan rate in lines per second (or lines per minute) (LR). The relationship between these parameters, the bandwidth and the resolution of the picture can be understood by examining *Figure 9-29* and *9-30*. Each square in *Figure 9-29* represents the area that is examined by the photodetector at the transmitter. This will correspond to an area at the receiver that will be reproduced to a corresponding level of gray. There are LPI rows of such spots scanned at both the receiver and the transmitter per inch of length LS of the source and copy documents. Similarly, there will be LPI columns of such squares in each inch of length L. Thus, each square inch of paper is covered by $(LPI)^2$ spots. The photodector/scanning system is then converting each square inch of the document into $(LPI)^2$ signal variations that electrically identify the document. These areas are referred to as picture elements or pixels or pels. If the paper is LS inches long and L inches wide, there are $LS \times L \times LPI^2$ pels per document.

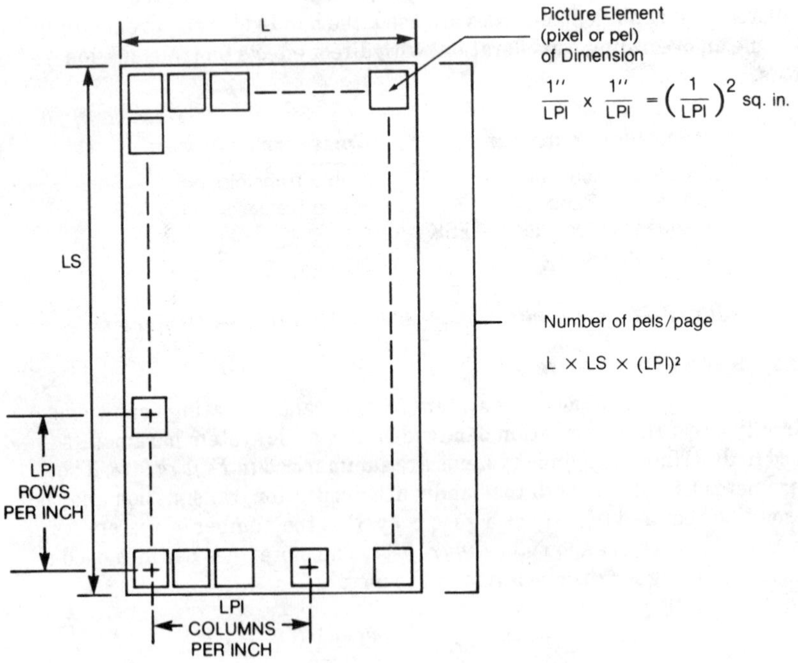

Figure 9-29. *Number of Picture Elements Per Page.*

An Example

In a typical facsimile scanning system there are 100 lines per inch (LPI = 100), and the length and width of the paper are 8½ inches and 11 inches, respectively (LS = 11, L = 8.5). In such a document the scanning system is examining 935,000 pels or picture elements. If the gray level in each pel were represented by a 6-bit digital code, which would distinguish between 64 different gray levels, the total number of binary bits required to represent the information in a document would be 6 times 935,000 or 5,610,000 bits. Suppose a communications system capable of sending 4800 bits/second is used to transmit this information. It would take about 20 minutes to transmit the entire page. For some applications this time would be acceptable; for others it would be much too long. If the resolution of 100 lines per inch is fixed, the only way to reduce the transmission time is to reduce the number of bits per pel or to increase the bit rate or both. These alternatives and their effect on system cost and performance will be discussed later in this section.

Bandwidth

Another way of looking at the communication system requirements is the bandwidth required for the transmission. *Figure 9-30* shows the maximum number of black bars that can be handled by the pel structure of *Figure 9-29*. Also shown in *Figure 9-30* is the voltage waveform that would appear at the output of the photodetector for such a bar pattern. These black voltage pulses would occur at a rate determined by the line scan rate LR and the line resolution LPI. For example, in a scan line of length L, inches would be L times LPI pixels. One half of the total pixels fall on bars that produce black level pulses in the electrical signal output of the detector. If the scan rate is LR lines per second, the time for a scan line is 1/LR seconds. Thus, $L \times LPI/2$ bars or pulses will occur in 1/LR seconds for a pulse frequency of $L \times LPI \times LR/2$ hertz.

If a system uses a line rate of 3 lines per second and a resolution of 100 lines per inch and a paper width of 8½ inches, the electrical signal would have a pulse frequency of $100 \times 8.5 \times 3/2 = 1275$ hertz. The communication system must be able to handle a maximum signal frequency of about 1300 hertz. If the LPI is increased or the scan rate is increased, the maximum bandwidth will increase proportionately.

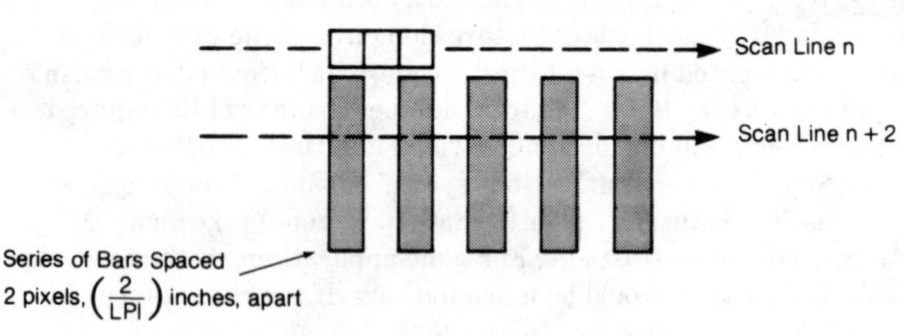

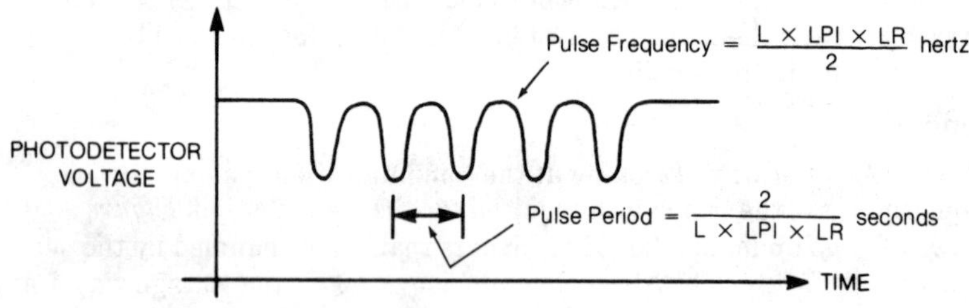

Figure 9-30. *Maximum Frequency Generated by Photodetector in a Facsimile System.*

The relationship between the signal bandwidth and the communications channel bandwidth depends on the type of modulation used. For example, the standard voiceband switched telephone network provides useful transmission for frequencies above 300 hertz and below 2500 hertz. In the example, if conventional amplitude modulation (double-sideband) is used, the communication system would have to pass at least twice the highest frequency of 1300 hertz for an overall bandwidth of 2600 hertz. This would not be a problem but might require a special, slightly better telephone line than the average which probably would cost more. A better approach would be to select a narrower band modulation approach such as single-side band AM. The single-side band approach would require only slightly more than the 1300 hertz bandwidth.

Another approach is to use a version of frequency modulation, frequency-shift keying or FSK. Typically a range of frequencies from 1500 hertz (for black) to 2300 hertz (for white) is used to encode the facsimile signal as shown in *Figure 9-31.* Readily available phase-lock-loop (PLL) integrated circuits can be used for the modulation process at the transmitter and the demodulation process at the receiver.

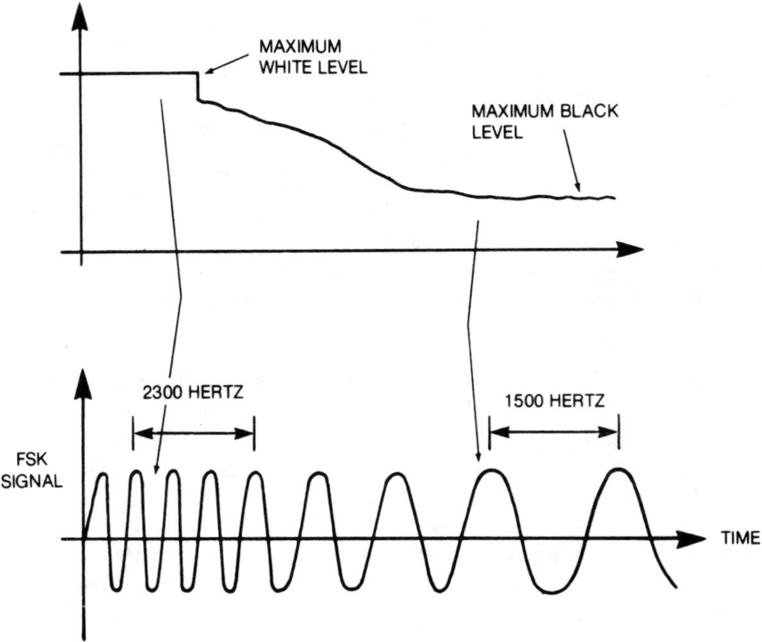

Figure 9-31. *Frequency Modulation of Facsimile Signal.*

Digital Transmission

When digital codes are used to represent the facsimile signal, it must be sampled at twice the highest signal frequency. The resulting output voltage sample must be converted to at least an equivalent 6-bit binary code. This code is then transmitted using AM or FM modulation on a leased telephone line or by using radio transmission. A standard dial telephone line can transmit at the rate of 4800 bits per second. If a picture is being scanned at the rate of 1 line per second and it has 800 pels per line, each requiring 6 bits per pel, 4800 bits/second would be required of the channel, which the standard telephone line can handle. However, to scan an entire page of 1100 such lines would require 1100 seconds of transmission time, which is about 18 minutes at long distance telephone rates.

If fewer bits per pel can be used, the number of bits per line can be decreased and the line rate increased. For example, if on the average 1 bit per pel was needed to provide an acceptable reproduction of the source document, the scan rate could be increased by a factor of 6 without increasing the bit rate requirements beyond that of an inexpensive telephone line. Yet the entire document could be sent in around 3 minutes. This process of encoding a pel with a small number of bits is called data compression. There are techniques for performing either bandwidth compression or data bit compression in communications systems. Some of these techniques are discussed in Chapter 8. As transmission costs increase, these techniques will become increasingly important in all communications systems, including facsimile transmission.

Higher Resolution

Thus far the effect of resolution of lines per inch on the bandwidth and data requirements of the transmission system has been summarized. While most standard printed and pictorial documents are scanned at about 100 lines/inch, other source material may require much higher resolution. Fingerprints require about twice this resolution to provide identifiable prints for law enforcement agencies. Newsphotos from wire services require resolution slightly above 100 lines per inch. Microfilm information requires more than 2000 lines per inch due to the highly reduced nature of microfilm information.

Synchronization

Line resolution is not the only parameter that effects the quality of the reproduced image. Just as in television systems, the received picture must be synchronized to the transmitted picture in order to avoid undesirable tearing or skewing when reproducing the copy. Thus, the scanning mechanism at the receiver must be "in-step" and synchronized with the scanning mechanism at the transmitter if acceptable results are to be obtained. This means that the receiver scan must start on the first line of the copy at the same time that the electrical signal from the first line of the source document comes from the transmitter. Further, during the reproduction process, the electronic circuitry of the receiver scanner must have stable characteristics, otherwise the reproduction will tend to drift continually to darker or lighter reproduction as the page is scanned.

If the scanning mechanism is operated in discrete movements instead of a continuous linear scan, these movements must not introduce jitter or distortion into the final reproduced document. All of these details are resolved in the final design of a high-quality facsimile system. Modern facsimile systems do offer good quality reproduction at reasonable speeds for a wide variety of source document types. In the future, the capabilities of facsimile systems should be even more impressive.

WHAT HAVE WE LEARNED?

- A facsimile system reproduces from a source document (printed page, typewritten page, pictures, etc.) a copy at a remote location using telegraph, telephone or radio transmission links.
- Facsimile systems are used to transmit business documents, weather information, law enforcement data, library reproductions and personal data.
- In a facsimile system there must be a device at the transmitter that converts the visual patterns of the source document into electrical signals for transmission.
- In a facsimile system there must be a device at the receiver that reproduces a copy of the source document from electrical signals received from the transmitter.
- Facsimile requires encoding optical information into electrical signals.
- Reproducing facsimile material can be done using electrostatic, electroresistive, and electrothermal techniques.
- Facsimile system characteristics of lines per inch resolution, lines per second scanning, picture element size and source document size help determine the type communications channel and the bandwidth of the channel used for transmission.

WHAT'S NEXT?

Worldwide communications are now almost as convenient as communicating with the neighbor next door. Much of this progress is due to communications by satellite – the subject of the next chapter.

Quiz for Chapter 9

1. A facsimile system differs from TV or computer print-outs in the following way:
   a. The facsimile produces a moving picture of the source image.
   b. The facsimile produces a typed output of a picture input.
   c. The facsimile reproduces the document at the transmitter onto paper or photographic film at the receiver.
2. Facsimile is similar to television in that:
   a. the original image is scanned in a raster pattern.
   b. the optical image is converted into electrical signals.
   c. the conversion of the optical pattern into electrical form occurs at high electronic speeds.
   d. the image is reproduced on the face of a TV picture tube.
   e. All of the above.
   f. a and b above.
3. The optical features of a document are converted to electrical form by having:
   a. a light spot illuminate a small portion of the document and a photodetector generate a voltage or current proportional to the reflected light.
   b. the document uniformly illuminated and have the photodetector detect the reflected light from only a small portion of the paper.
   c. Either of the above.
4. If a facsimile machine is designed to provide a resolution of 200 lines per inch, how many picture elements are in an 8½″ × 11″ page?
   a. Less than 1 million.
   b. 1 million to 3 million.
   c. Over 3 million.
5. If the receiver is a cylinder rotating at 180 rpm in the system of question 4, how many minutes will it take to reproduce the entire page?
   a. Less than 5 minutes.
   b. 5 to 10 minutes.
   c. Over 10 minutes.
6. What bandwidth would be required to send the electrical signal of the system of 4 and 5?
   a. 1275 hertz.
   b. 2025 hertz.
   c. 2550 hertz.
   d. 3125 hertz.
7. Match the system features with the printing method:

a. electroresistive	A. low cost
b. electrothermal	B. high speed
c. impact	C. high contrast
d. ink jet	D. good gray level reproduction
e. electrostatic	E. multiple copy

8. In a cylindrical scanning system:
   a. the paper being scanned is wrapped around a cylinder.
   b. the cylinder is rotating while the scanning optoelectronics is slowly moved down the paper.
   c. the paper is loaded horizontally into the system.
   d. a wide range of documents and paper weights can be used.
   e. All of the above.
   f. a and b above.
9. A facsimile receiver can be used to expose photographic plates if:
   a. the electrical current is caused to flow through the photosensitive film.
   b. the electrical facsimile signal is converted to a scanning light beam of varying intensity which is focused onto the film.
   c. the electrical signal is converted to a charge pattern on the film.
10. In the system of problems 4 and 5, if the information is to be sent using an 8-bit PCM, how many bits per second are required?
    a. Less than 30,000.
    b. 30,000 to 40,000.
    c. 40,000 to 50,000.
    d. Above 50,000.

(Answers in back of the book)

Satellite Communications Systems

ABOUT THIS CHAPTER

The last several chapters have concentrated on specific types of communications, from telephone systems to facsimile. In this chapter the delivery of these various types of information will again be the principal concern. Throughout history the communication task has consisted of two problems. First the information must be encoded into a form that can be transmitted and that will be understandable to both the source (transmitter) and the destination (receiver) parties. Secondly, the information must be carried from the source location to the destination location. Electronic communications solves the first problem by converting the source information into electronic signals and reconverting these signals into a suitable form at the receiver. The techniques used have been covered in the past four chapters for various types of information, including sound (speech, music, etc.), pictures (still documents and moving action scenes), and computer data. Delivery of the information over transmission links of wires and cables has been discussed in some detail. The general features of electromagnetic radiation delivery have also been discussed. In this chapter, the most modern method of information delivery using electromagnetic radiation will be covered, that of the satellite relay system.

WHAT ARE THE BASIC FEATURES OF SATELLITE COMMUNICATIONS?

To define what satellite systems are, one must look at the basic features and historical development of these systems. Satellite systems are a relatively new development in the history of communications since their use had to await man's entry into the space age in the late 1950's. Satellite systems depend heavily on technology, including rocketry, space mechanics, solid-state electronics, high-frequency electronics and radiation, and modern communications networks. Some of the problems that have to be solved to insure a successful satellite system can be understood by examining the basic features of such a system.

Basic Satellite System

Figure 10-1 illustrates the basic components of a typical satellite system. The satellite is orbiting around the earth and receives its energy from the sun through solar cells. It has one or more antennas which receive radiation from the earth and send radiation back to the earth. One earth station (point A in *Figure 10-1*) transmits information to the satellite at a specified carrier frequency, typically in the 6 billion cycles per second or 6-gigahertz (abbreviated 6 GHz) band. This is called the up-link frequency. The satellite receives this transmitted radiation and information, repeats and reinforces it by transmitting it to the earth on a different carrier frequency, typically 4 GHz. This is called the down-link frequency. The satellite is said to be operating in the 6/4 GHz frequency bands.

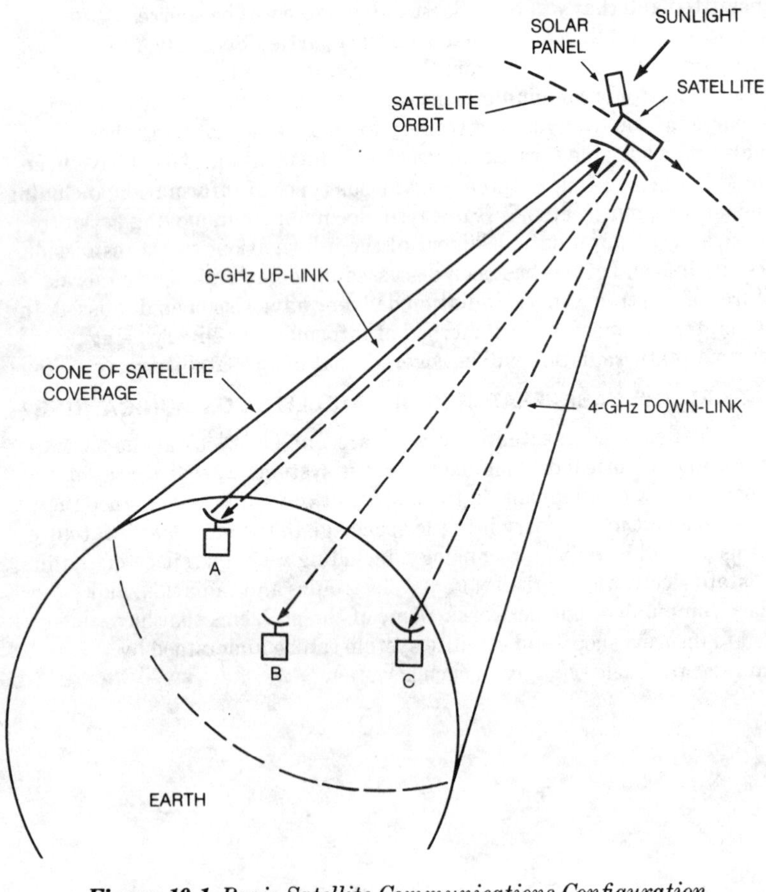

Figure 10-1. *Basic Satellite Communications Configuration.*

In the case of *Figure 10-1*, the satellite antenna has been designed to provide radiation to all parts of the earth visible to the satellite. For a satellite operating at a distance of 22,000 miles (35,800 km) from the earth approximately 40% of the earth's surface would be exposed to this radiation. By using a single satellite in such an orbit, earth station A can send information to any other earth station, including itself, within this 40% area of the earth's surface. If the information is to be sent from point A to point B, nothing prevents point C from receiving the information, unless the information were coded in such a way that station C could not understand it without a proper key. In fact, in some applications the station at point A may want to send information to stations C and B as well as a large number of other stations over the earth's surface. Distribution of TV programs to cities and towns on the same or different continents from a single central studio is one example of such broadcast systems, and for this the satellite becomes a broadcast satellite.

Station Conditions

Two situations occur at the receiving station. First, the information sent may be received and understood by all stations but of interest to only one (or several). Secondly, the information may be intended to be received and understood by only one station. For example in *Figure 10-1*, if station B and C receive and understand the transmission from station A, and station B is the only one interested, station C ignores the information and doesn't respond. However, if the information must be received and understood only by station C, the information must be encoded in such a way that station B would not understand it even if it "listened in" or attempted to steal or pirate the information. Such encoding for the purposes of security (cryptology) was covered in the last chapter.

One other important point about the satellite broadcast signal over the world is that station A can receive its own information and check it for errors. If it is very important that the information be sent without error, this can be an important tool for making sure that the information was correct when it left the satellite. Many techniques exist that can be used at the receiving station for detecting and correcting errors that occur during the transmission of the information. In the last chapter a basic example of one of these techniques was discussed.

Directional Radiation

One other feature illustrated in *Figure 10-1* is that the transmitting earth station must provide a very directional beam of radiation that will be received by the satellite being used and not some nearby satellite, because there are many satellites in orbit. As a consequence, the earth station transmitter and antenna must be carefully designed to achieve the proper beam width. It must also be able to locate and track or follow the position of the satellite, so that its beam will not miss the satellite antenna. This problem is simplified if the satellite stays in the same position relative to the earth at all times. Since the earth is spinning on its axis, the only way to achieve a satellite position fixed relative to the earth's surface is to have the satellite motion match that of the earth's rotation. An orbit that achieves this match is called a geosynchronous orbit.

Geosynchronous Satellite Orbits

To be in a geosynchronous orbit the satellite must be in an orbit 35,800 kilometers or about 22,000 miles from the center of the earth. It must further be in a plane that contains the equator. This type of orbit is illustrated with a top view in *Figure 10-2a*. In such an orbit, if the earth station's position changes from point A to point B, the satellite moves from point C to point D. A line drawn from the center of the earth through point A hits the satellite at point C, and similarly a line from the earth's center through point B hits point D, so that the earth station's position has not changed relative to the satellite position. If the earth station's antenna is pointing toward the satellite's antenna at point A, it should still be pointing at the satellite's antenna at point B. While minor adjustments in the position of the earth's antenna may be required from time to time it will be able to remain locked on the satellite relatively easily.

Other Orbits

Figure 10-2b shows what happens if the satellite is in a lower orbit. In this orbit the satellite moves faster. As a result, a considerable change in the earth station's antenna position is required to remain locked on the satellite at point D, while the earth station rotates from A to B. Eventually, the satellite will race past the earth's horizon, causing the earth's station to no longer have access to the satellite at point F when the earth station has moved to point E. Unless another satellite in the same orbit comes into view as the previous satellite disappears below the horizon, the earth station would not be able to use satellite communication more than about 8 hours a day. Even with several satellites in the orbit, the portion of the earth's surface covered at any given time would be continually changing and the earth station antenna would have to be able to cover the entire arc of the sky. This would require a very expensive antenna control system. As a result of these problems, modern satellites use the geosynchronous orbit if it is available.

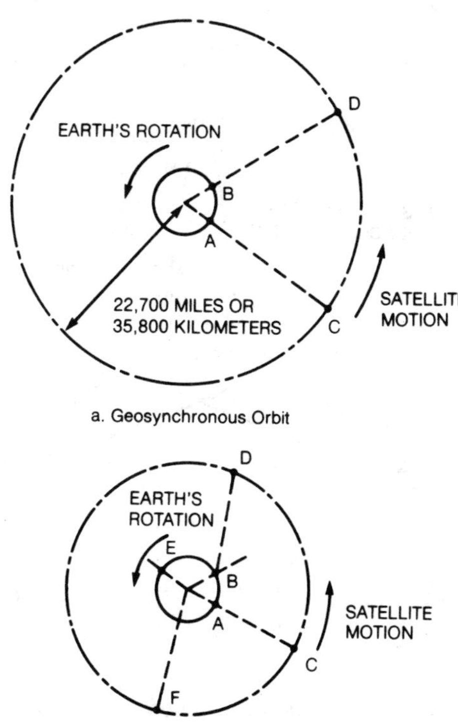

Figure 10-2. *Satellite Orbits.*

Equally Spaced Satellites

Even with geosynchronous orbits, the satellite's coverage of the earth's surface is limited to about 40%. However, three such satellites spaced at 120-degree intervals, as shown in *Figure 10-3,* could cover all of the civilized world with the exception of some polar areas. Of course, only earth stations in the coverage of two satellites could communicate from one zone to another. Such international stations would be able to relay messages or communication from one zone to another for complete world coverage. Stations contained within a single zone can only communicate with stations within that zone and are thus considered local stations.

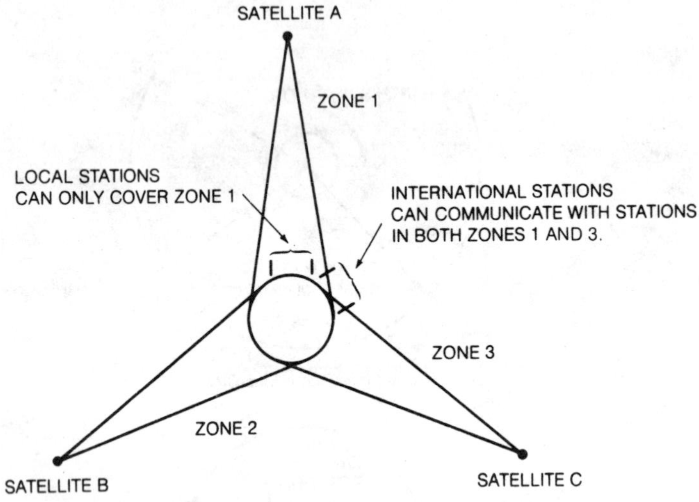

Figure 10-3. *Coverage of Earth's Surface with Three Geosynchronous Orbiting Satellites.*

Satellite Channels and Station-to-Station Communications

For stations within the zone of a given satellite, shown in *Figure 10-4,* any number of independent station pairs can communicate with each other. While the satellite is broadcasting to the entire visible portion of the earth's surface, each station listens only to the signal intended for it. For example, communication can be occurring between stations A and B (along a signal path #1) at the same time communication is occurring between stations C and D (along a signal path #2). Upon first examination this doesn't seem possible since the antenna is receiving signal frequencies in the 6-GHz range from both stations A and C at the same time. In addition, within the same time period, the satellite is broadcasting to the entire visible area below on carrier frequencies in the 4-GHz range. Obviously, some multiplexing must be used to allow these communications to occur simultaneously.

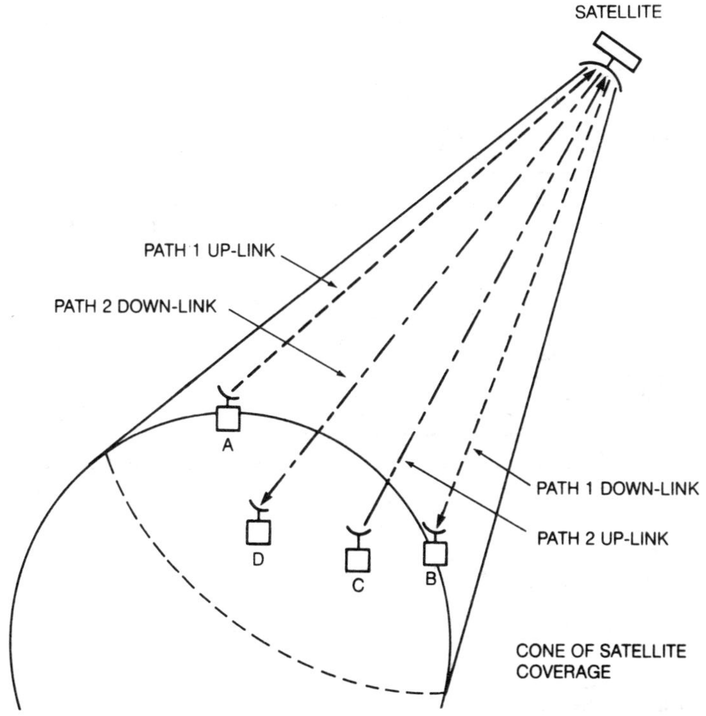

Figure 10-4. *Multiple Communications Using Same Satellite.*

Multiplexing Satellite Signals

There are two general approaches used to achieve this sharing of a satellite: frequency division multiplexing (FDM) and time division multiplexing (TDM). Most satellites to date have used some variation of the FDM approach, though more and more of the TDM approach will be used in the future. A typical FDM scheme is shown in *Figure 10-5.* Each satellite contains a certain number of transponders, which are a receiver-transmitter pair. The receiver of a satellite is a wideband receiver that covers the entire range of up-link frequencies shown in *Figure 10-5a.* These frequencies are typically from 5.925 GHz to an upper frequency limit that depends on the number of channels the satellite handles. In *Figure 10-5a,* which is the FDM structure for the 12-channel Western Union satellite, each channel has a bandwith of 36 MHz with a 4-MHz spacing between channels. Above channel 12 is a 20-MHz command and control or telemetry channel. Thus, the receiver frequency band extends from 5.925 GHz to 6.425 GHz. The transponders in the satellite convert this frequency range to the down-link channel frequencies shown in *Figure 10-5b,* which cover the 3.7- to 4.2-GHz band.

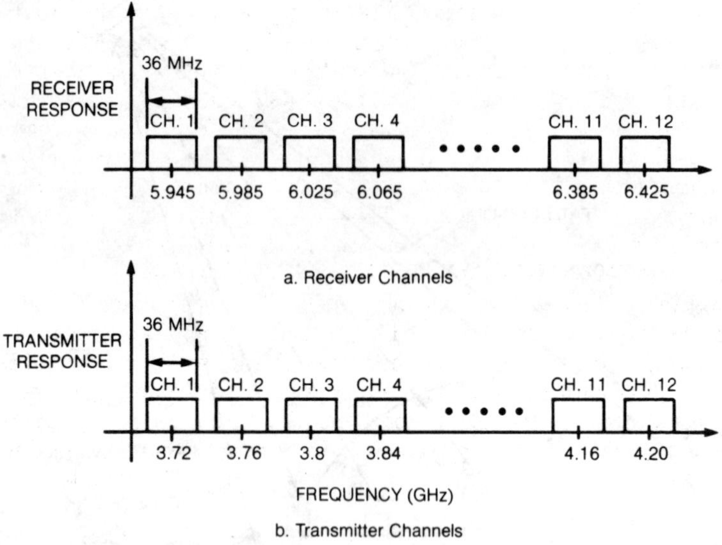

Figure 10-5. *Frequency Division Multiplexing in a Typical Satellite.*

Channel Assignments

In order for a pair of earth stations to communicate via the satellite, one of the channels will have to be assigned to the communication task. Thus, in *Figure 10-4,* the A-B communication could be assigned to channel 1 and the C-D communication could be assigned to channel 2. In this way, the transmissions do not interfere with each other and can occur simultaneously because different frequencies are being used by the A-B and C-D station pairs. Obviously, station A must know that channel 1 is available and station B must know that A is transmitting to it over channel 1. Station B then can accept the information coming in on channel 1 and reject any information coming in over the other channels. A similar set of conditions applies to the C-D communication over channel 2.

These channel assignments could be made permanently, in which case, if station B detects information on channel 1 it is information being sent to it by station A. If it detects information on channel 2, it would know it was information intended for station D. Of course, this would be a wasteful approach, since the A-B communication may not be continuous for 24 hours a day. For those periods that A was not communicating with B, channel 1 would be idle when it could have been used by other stations.

FDMA

A more reasonable approach is to assign available channels to the next station requesting or demanding service, and notify the destination station which channel is being used. Then, once the communication is over, that channel would be released back to the available pool. This approach is called Frequency Division Multiple Access or FDMA. The command channel can be used to control the allocation of channels, and to notify the transmitting station what channel is available and to notify the receiving station which channel is being used. This control function can be handled by a central computer or by computers at each earth station. This same strategy would work well if station A is a broadcast station. The transmitter, station A in this example, would request a channel and state that transmission was to be received by stations B, C, and D. The control system would assign the available channel and notify stations B, C, and D to receive information from station A on the assigned channel. With the broadcast and station-to-station capability, the satellite system has all the features required of any information delivery system. It further does this without the need for an extensive repeater-relay system between the communicating points.

General System Comparisons

All transmission systems prior to the satellite system, with the exception of radio transmission, required a network of wired interconnections between points A, B, C, and D of *Figure 10-4* (and all other stations to be communicated with). This is one of the advantages that satellite systems have over alternative systems as shown in *Figure 10-6*. All but the radio link and satellite link require repeaters every 2 to 40 kilometers. This may not be feasible when the transmission must span oceans or access remote areas. In some of these cases the satellite or radio link is the only possible solution and certainly the most economical solution. Of these, the radio link is somewhat unreliable since transmissions over the surface of the earth have noticeable electromagnetic interference from other signals and fading of signals due to ionospheric variations. In addition, the data rate of radio links for worldwide communications is considerably lower than the satellite system. For high-data rates the bandwidth requires a microwave high-frequency carrier, therefore, microwave waveguide or radio links or satellite links must be used.

System Parameter	Satellite	Cable/Wire	Microwave	Radio
Repeater Spacing (km)	No Repeater	2-40	1-40	Depends on frequency
Weather Interference	Low	Low	Low	High
Bandwidth	Very High	Low-Medium	Very High	High
Security	Poor	Fair	Poor	Poor

Figure 10-6. *Alternative Transmission System Comparison.*

Cryptology must be used to make any of the systems secure, otherwise the transmissions are relatively open to anyone that really wants to obtain them. Cable and wire systems rate higher for security than the others. *Figure 10-6* seems to imply that satellite communications systems should dominate. There are several reasons not listed in *Figure 10-6* that prevent this from happening at the present time. The most important of these is cost. Present systems are very expensive, but what is important is that costs per channel have decreased significantly over the last 20 years and the trend is continuing.

HOW HAS SATELLITE SYSTEM CAPACITY EVOLVED?

In order to make satellite communications economically competitive, the cost per communication channel must be minimized. Reducing this cost for a given satellite configuration requires that the maximum number of channels possible be placed on a carrier. Therefore, the maximum bandwidth that can be provided by state-of-the-art technology must be used. The technologies that impact both satellite costs and capacity include: launch vehicle technology, solid-state device and system technology, and antenna technology.

Rocket Launch Technology

The amount of satellite weight that can be placed in geosynchronous orbit depends on the rocket used to launch the satellite. The larger the rocket, the more weight that can be injected into orbit, and the higher the cost of the launch. The more weight available for the satellite, the more power and channel capacity it can have. These two factors are illustrated in *Figure 10-7*. *Figure 10-7a* is a plot of the launch cost for a given satellite orbit weight. Only three rockets are shown, but the trend of higher launch cost for higher orbit weight is evident.

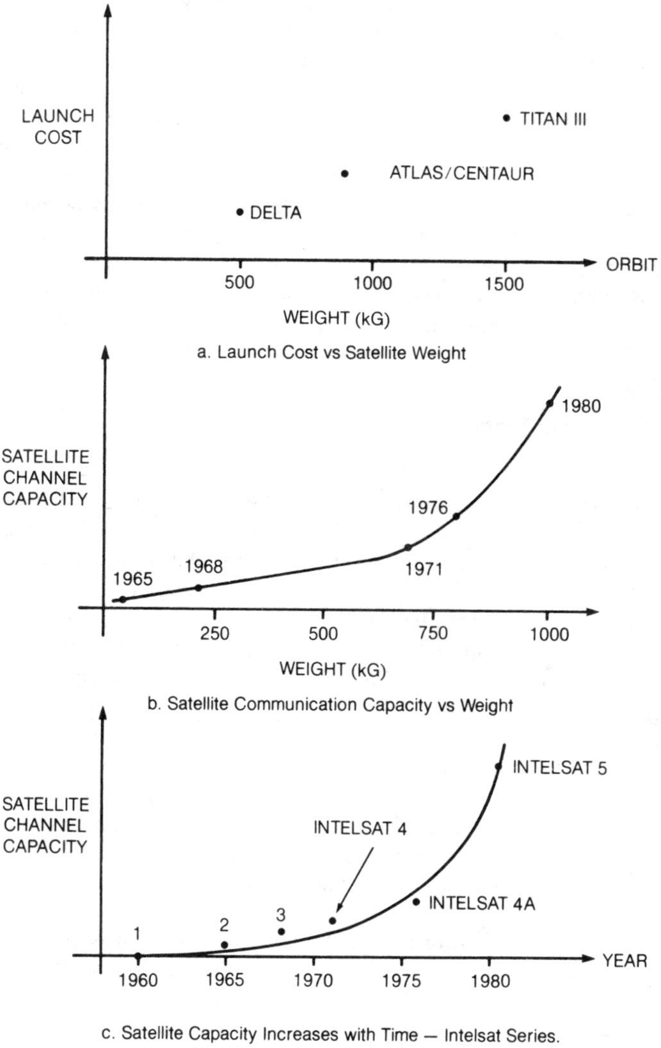

LAUNCH COST

• TITAN III

• ATLAS/CENTAUR

• DELTA

ORBIT

500 1000 1500

WEIGHT (kG)

a. Launch Cost vs Satellite Weight

SATELLITE CHANNEL CAPACITY

1980

1976

1971

1965

1968

250 500 750 1000

WEIGHT (kG)

b. Satellite Communication Capacity vs Weight

SATELLITE CHANNEL CAPACITY

INTELSAT 5

INTELSAT 4

INTELSAT 4A

1 2 3

YEAR

1960 1965 1970 1975 1980

c. Satellite Capacity Increases with Time — Intelsat Series.

Figure 10-7. *Increases in Satellite Channel Capacity.*

The other factor illustrated in *Figure 10-7a* is that a certain rocket will launch a certain maximum weight into orbit, so that the satellites are designed to match these discrete weight options. The DELTA series of rockets will launch into orbit a weight of 500 kg, the ATLAS/CENTAUR rocket a 900-kg weight and the Titan rocket a 1500-kg orbit weight. However, as *Figure 10-7b* shows, these heavier satellites have had a much higher capacity, with the rate of increase of capacity versus weight much higher than the increased costs due to weight. Some of this capacity increase has definitely come from advances in electronic technology since the heavier satellites have been launched at later dates. *Figure 10-7c* summarizes the information. It shows the increase in capacity by year. It is quite clear that improvements in technology and the ability to place more weight into orbit have dramatically increased the channel capacity of a satellite. But launch costs are only part of the total system costs. There is the cost of the satellite and the cost of the earth stations that have to be considered, and their cost reduction depends on advances in electronic and antenna technology.

Satellite Electronic Technology

Solar Panels

One area that depends on modern solid-state devices and that impacts the satellite capability is the solar cell energy source. The amount of power generated from the sun's energy depends on the total area of the solar panels that is exposed to the sun and the generation efficiency of the cell. Significant improvement has been made in solar cell efficiencies since present panels generate more than 50% more power than early units for the same panel area and weight. Increased power can be used to increase the RF energy level radiated from the satellite, which can reduce the costs of earth station components. The extra power can also be used to add additional channels to the satellite and increase its communication capacity.

Two types of solar cell mountings have been used for satellites. *Figure 10-8a* shows a spin-stabilized satellite. The solar cells are mounted on a cylindrical body and continually rotated so they face the sun while the antenna is kept pointed at the earth. The spinning cylinder acts as a gyroscope to keep the satellite oriented in space. The spin-stabilized satellite only has about half of the solar cell weight that is useful for communications since only 40 to 50% of its solar cells are illuminated by the sun.

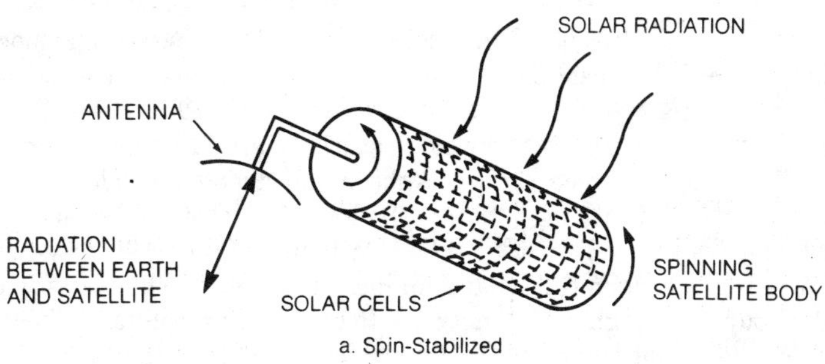

a. Spin-Stabilized

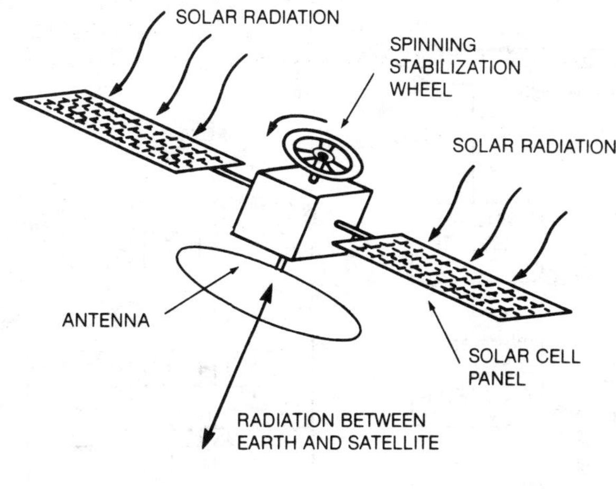

b. Body-Stabilized

Figure 10-8. *Types of Satellites.*

Improvements were made by using the body-stabilized satellite of *Figure 10-8b.* Here a spinning wheel provides the gyroscope and the solar panels and body of the satellite are fixed. The entire solar panel surface is illuminated by the sun at all times and much larger areas than just the body area can be used. The satellite power is increased significantly over the spin-stabilized satellites. For example, recent body-stabilized satellites such as CTS, Intelsat 5 and ATS units generate kilowatts of power while, early spin-stabilized units generated 40 watts (1965) to 400 watts (1971).

Electronic Circuitry

The electronics that are used in the receiver, filters, and low-power portions of the satellite communications systems have also become more weight efficient in the past 20 years of satellite development. The basic block diagram of a typical satellite is shown in *Figure 10-9*. The 6-GHz antenna receives signals from the earth stations, including any command signals. The wideband receiver boosts the signal strength to a level where it can be converted, by mixing with a 2-GHz local oscillator, to the 4-GHz range and drive a traveling wave tube (TWT) preamplifier. A traveling wave tube is an electron tube that amplifies microwave frequencies. This signal is routed through a filter network to the power amplifier TWT that generates the signal for the 4-GHz transmitting antenna for retransmission back to earth. A filter is a bandpass amplifier.

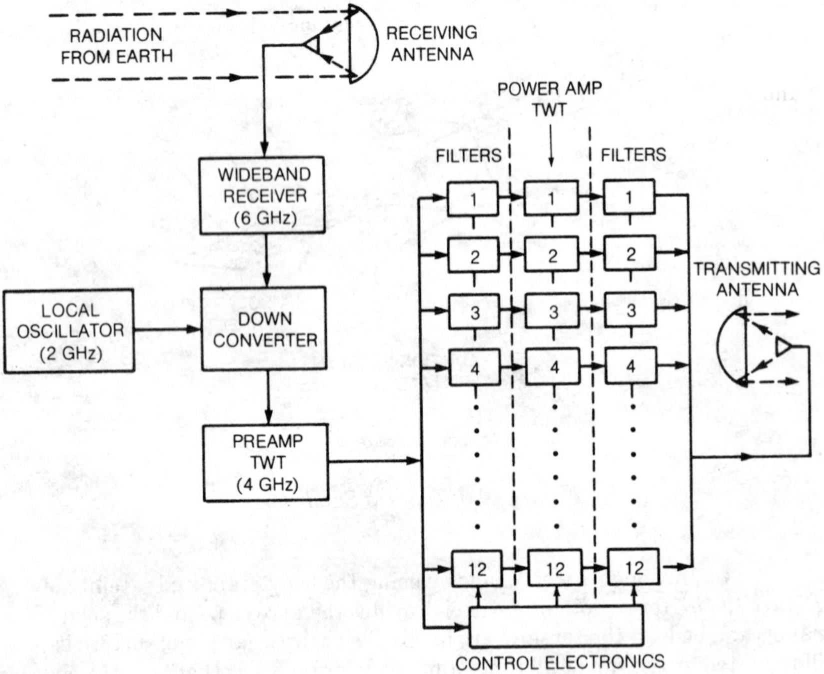

RADIATION FROM EARTH
RECEIVING ANTENNA
POWER AMP TWT
FILTERS
FILTERS
WIDEBAND RECEIVER (6 GHz)
LOCAL OSCILLATOR (2 GHz)
DOWN CONVERTER
PREAMP TWT (4 GHz)
1 2 3 4 12
TRANSMITTING ANTENNA
CONTROL ELECTRONICS

Figure 10-9. *Typical Satellite Receiver/Transmitter Structure.*

By making the wideband receiver, down converter and filters out of microwave integrated circuits and using modern weight efficient traveling wave tubes, it is possible to maximize the number of transponders (transmission channels) available on a satellite of a given orbit weight. The particular channel used is controlled by the signals sent over the control channel. Even with a reasonable size solar power generator and modern power and weight efficient electronics, the overall satellite capacity and performance depends on delivering the RF or electromagnetic energy back to earth. This depends on the antennas used in the satellite and at the earth station.

Antenna Technology

One way to get the most out of the available satellite power is to design the antenna so that it delivers a signal to the earth areas that need it, and avoids sending to areas which contain no earth stations. Generally, the larger the antenna, the more precisely focused the radiation beam. The more precisely the radiation pattern is controlled, the closer satellites can be spaced in orbit without interfering with one another. Thus, the total bandwidth available from the geosynchronous orbit can be increased by using larger, more carefully designed antennas. Also, since the power or energy hitting the earth's surface is increased by a better focused antenna, smaller earth station antennas are needed. This can significantly reduce the cost of earth stations, while moderately increasing the cost of the satellite.

Currently, satellite antennas of about 30 feet (9m) in diameter have been used successfully in space with the ATS series of satellites. The problem with such large antennas is that special techniques are required to couple the energy from the power amplifier to the antenna. At microwave frequencies, energy comes from the power amplifier to the antenna through waveguides. The end terminator on the waveguide that feeds energy to the antenna is called a "horn." As shown in *Figure 10-10*, several feed horns must be used with the large antenna. Even though this increases the weight of the satellite it has operational advantages. For example, if each feed horn corresponds to a particular area on the earth's surface, as illustrated in *Figure 10-11*, then if the satellite is only transmitting from area 1 to area 2, horns can be selected so the power of the satellite can be delivered only to area 2, allowing earth stations with small diameter receiving antennas and inexpensive receivers to receive the information.

Another approach is to use separate antennas for global coverage and localized spot coverage to allow two different transmission paths to use the same frequency channel for separate communications. Such frequency re-use is a feature that will be common in the future. These techniques have the effect of dramatically increasing satellite capacity without a significant increase in satellite costs.

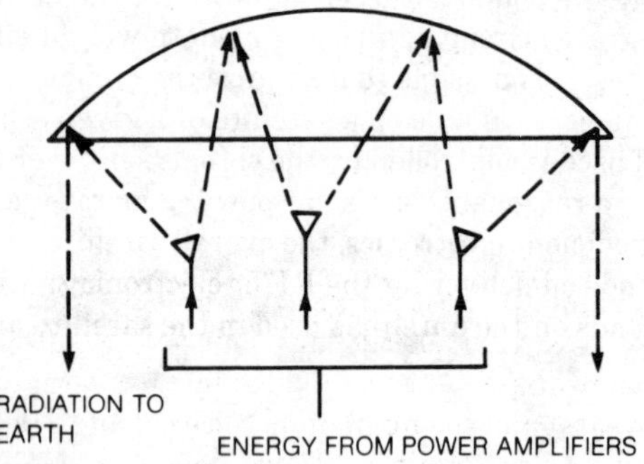

Figure 10-10. *Feed Assemblies for Large Satellite Antennas.*

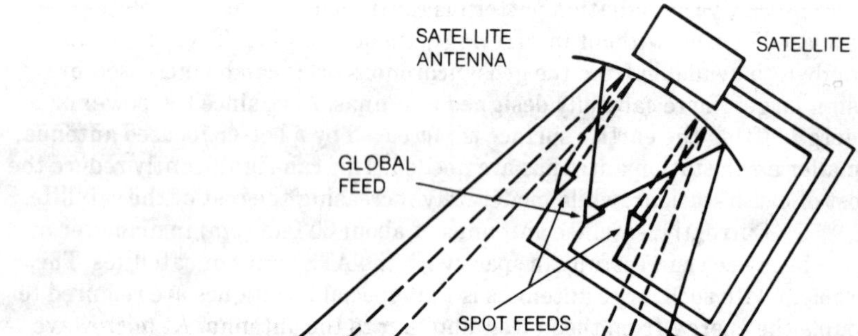

Figure 10-11. *Multiple Feed Antennas and Radiation Patterns.*

Summary of Satellite Evolution and Costs

The Intelsat series of satellites provides a case-study of the advances that have been made in the design of typical satellite systems. In fact, *Figures 10-7b* and *10-7c* were based on Intelsat features. *Figure 10-12* summarizes the changes made in the capabilities of this series of satellites from 1965 to the present. *Figure 10-13* summarizes the effect of technology improvement on the cost per voice channel for this series of satellites. According to Bargellini[1], the cost per voice channel per year of the Intelsat 4A launched in 1971 is around 1000 dollars. This is total satellite cost, including launch and equipment cost, divided by the average expected life of the satellite (7 years) and the number of voice channels (6000). The total satellite cost for an Intelsat 4 unit in 1971 was of the order 30 to 40 million dollars.

System Parameter	1965-68 Intelsat 1-3	1971 Intelsat 4	1976 Intelsat 4A	1980 Intelsat 5
Voice Channels	240-1200	4000	6000	12,500
Bandwidth (MHz)	50-130	500	800	2300
Number of Transponders	1-2	12	20	27
Transponder Bandwidth (MHz)	25	36	36	40,80,240
Earth Station Antenna Diameter (Meters)	25.9 (85′)	30	30	30,10
Satellite Weight (kg)	38-152	700	790	950

Figure 10-12. *Satellite Evolution – Intelsat Series – 1965-1980.*

The ground station costs must be added to these yearly costs for the total cost for each user. Much of the ground station costs depend on the size of antenna required. As shown in *Figure 10-12*, the early 1965 stations used an 85-foot (25.9m) antenna while current Intelsat earth stations use a 30-meter (97 foot) diameter antenna for Class A stations or a 10-meter antenna for Class B stations. When the satellite is sending to a Class B antenna station it must increase its radiated power to maintain good communication quality and it charges more for such customers. However, such customers do benefit from less expensive earth stations and smaller installations.

[1]Bargellini, P.L., "Commercial U.S. Satellites," pps. 30-37, Vol. 16 No. 10 Oct., 1979, IEEE Spectrum.

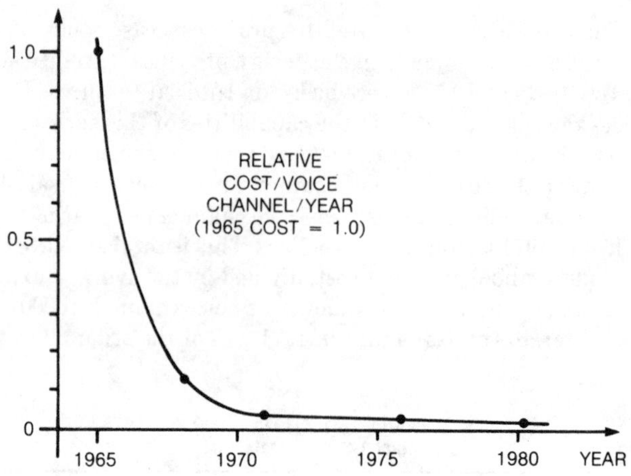

Figure 10-13. *Effect of Technological Improvements on Voice Channel Costs.*[2]

From *Figure 10-12* it can be seen that 1980 satellite capabilities include up to 27 transponders which provide up to 12,500 channels that may be used for voice channels or for computer data at various data rates. Total bandwidth has been increased to 2300 MHz. Generally, a given transponder in a satellite can be allocated as needed to a given communication task. For example, the 36-MHz bandwidth transponder could be used to send a TV channel, 1200 voice channels, or data at a rate of 24 megabits per second.

Geo-Satellite Position

At the current time, the satellites in geosynchronous orbit are placed as illustrated in *Figure 10-14*, with the heaviest concentration being over the U.S. and the Atlantic Ocean (for trans-Atlantic communication). The orbit is really not very crowded at the present time. A satellite of 500-MHz bandwidth can be spaced at 2 or more degrees from its neighbor without interfering with the neighbor as long as ground antennas in the 30-meter range are used. As the number of geosynchronous satellites are increased or the diameter of earth station antennas are decreased (meaning increased power output), the orbit could become saturated. Thus, the search continues for increased channel capacity and functional capability for each satellite.

[2]Martin, James, "Future Developments in Telecommunications," pp. 222, Prentice-Hall, 1977.

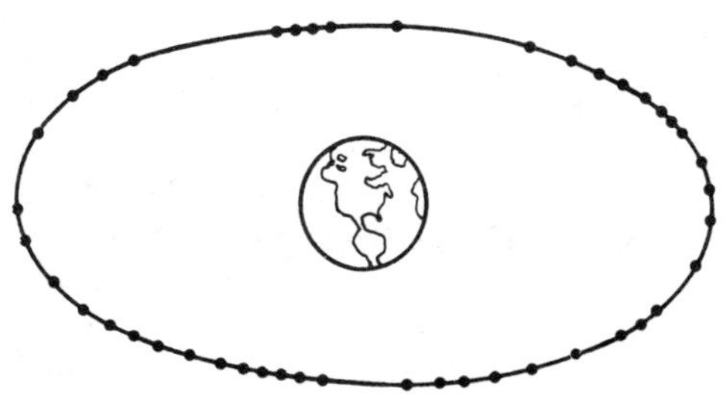

Figure 10-14. *Approximate Distribution of Geosynchronous Satellites. (1980)*[3]

HOW WILL SATELLITE CAPACITY BE IMPROVED IN THE FUTURE?

Every technique that is feasible for increasing satellite capacity and decreasing channel costs to customers will be pursued in the future. Development continues in these areas: more efficient networks, improved switching, increased multiplexing, more efficient and smaller antennas (especially at earth stations), and more powerful beams for transmission links. More and more efforts will be devoted to allowing smaller antennas at the earth stations. Specialized transmission techniques using polarized radiation or multiple bands will come into use. Let's look at a few of these techniques.

Radiation Techniques

One approach to reducing the channel cost of satellite capacity is to increase the power that is received at the earth station. At a high enough power level, the earth station antenna can be made smaller, the receiving station electronics reduced and the total cost can be lowered to reasonable levels. The noise characteristics, the focus and the directivity of the earth station transmitted beam are affected by the size of the antenna. The bandwidth required to frequency modulate the carrier to achieve reliable transmission is directly related to intersatellite noise characteristics.

[3]Edelson, B. I., "Satellite Communications In The Next Decade," pp. 60, *Satellite Communications Technology,* 14th Goddard Memorial Symposium, Univelt Inc., 1977.

Earth Station Antenna Size

The effect of earth antenna size on allowable satellite spacing to avoid interfering with adjacent satellites is shown in *Figure 10-15.* Considering that the minimum allowable spacing is used, the earth station antenna would have to be at least 100 wavelengths in diameter. Consequently, holding the same beam width, the higher the carrier frequency the shorter the wavelength and the smaller the permissible antenna diameter. For example, a 6-GHz up-link carrier has a wavelength of 5 centimeters. As a result, the antenna would have to be above 5 meters (or about 15 feet) in diameter. This is not excessive for an antenna on top of a large building or on the grounds of a large business plant, but it would be excessive for home use or mobile use. If now the carrier frequency is changed to 15 GHz, with a wavelength of 2 centimeters, a 2-meter (or about a 6½ foot) diameter antenna could be used. Increasing the frequency further to 30 GHz would allow the diameter to be reduced to 1 meter, or about 3.3 feet, which is quite feasible for home or mobile use.

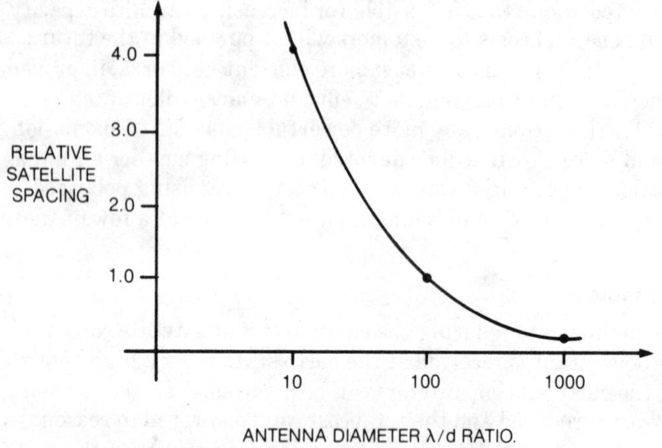

Figure 10-15. *Relationship Between Antenna Diameter and Satellite Orbit Spacing.*[4]

[4]Jansky, Donald, "Tools for Quantitative Analysis," p. 110, Communications Satellite Systems, 1975, IEEE Press.

Noise

Noise in an antenna used only for receiving signals varies with the antenna beam width. *Figure 10-16* illustrates the problem. If the beam width of the receiving antenna includes other satellites so that it detects a wide area of stray electromagnetic radiation, all this signal appears at the receiver as noise and tends to block out the desired (Satellite #1) signal. Desired signals buried in noise in this fashion require a very wide-band FM transmission to effectively detect the signal. Because small antennas with 1-meter to 3-meter diameters cannot provide as narrow a beam as larger antennas, up to 10 times normal bandwidth may be required. Thus, a 4-MHz TV signal would take the entire 36-MHz bandwidth of a satellite transponder to reliably transmit a high quality picture. Similarly, a 4-kHz voice channel might require a 40-kHz bandwidth FM signal to properly transmit it. This would mean that only 100 such voice channels could be sent on a single transponder instead of 1000. The earth station cost is reduced by using a smaller antenna, but as a result the transmission cost is increased significantly because the satellite bandwidth is being used inefficiently. Motivated by the need to service smaller and smaller earth stations, satellite frequencies continue to increase because system penalties seem to be minimized at the higher frequencies. *Figure 10-17* shows frequencies in use or planned for satellite systems. The 6/4-GHz C band is the one used most in the past. Increased use of the 14/12-GHz Ku band will continue, especially for inexpensive earth stations, until that band becomes as heavily subscribed as the current C band. Then, the K band use will increase.

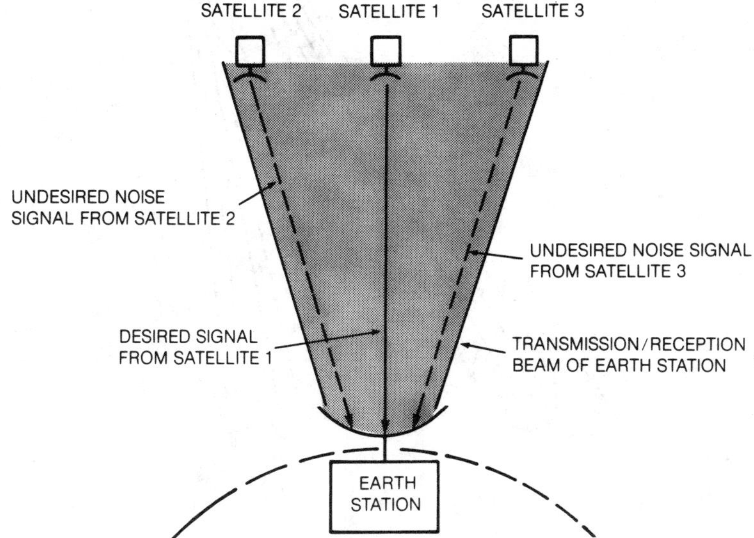

Figure 10-16. *Effect of Earth Station Antenna Beam Width on Signal Interference.*

Band	Up-Link Frequency GHz	Down-Link Frequency GHz
C	6	4
Ku	14	11/12
K	29/30	19/20

Figure 10-17. *Frequency Bands Used for Satellite Communications.*

Spot and Polarized Beams

A radiation technique that should increase in use was mentioned previously when *Figure 10-11* was discussed. The technique is referred to as "spot beaming" and its effect is shown in *Figure 10-18*. Stations A and B in the west spot can use the same frequencies as are used by stations C and D in the east spot. They just must be multiplexed properly to keep from interfering with each other. Proper switching of transponders can be provided to accomplish this.

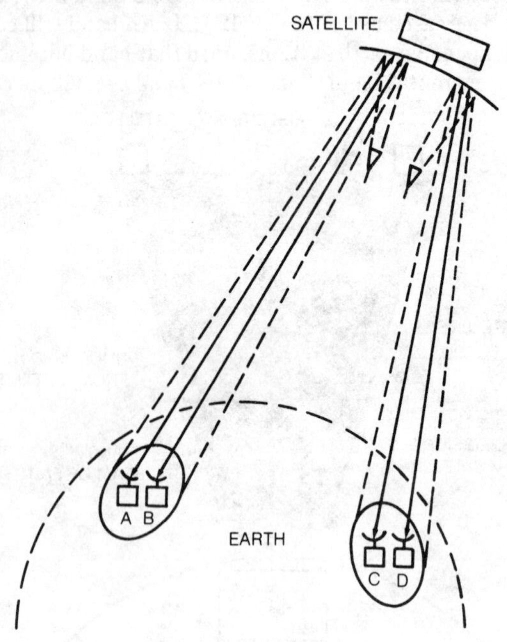

Figure 10-18. *Re-Use of Frequencies Through the Use of Spot Beams.*

Another approach to allow the satellites to use the same frequencies in parallel without interference is through the use of polarized radiation. The system setup is shown in *Figure 10-19*. One pair of satellite antennas is vertically polarized, and another pair is horizontally polarized. Either vertically or horizontally polarized transmissions are received by the respective antenna and retransmitted in the same polarization. Filter #1 may be 3.7 to 3.74 GHz for the vertical polarization path, and Filter #2 may be 3.72 to 3.76 GHz without channel interference. This scheme was used in the COMSAT satellite and the Intelsat units from 4A on.

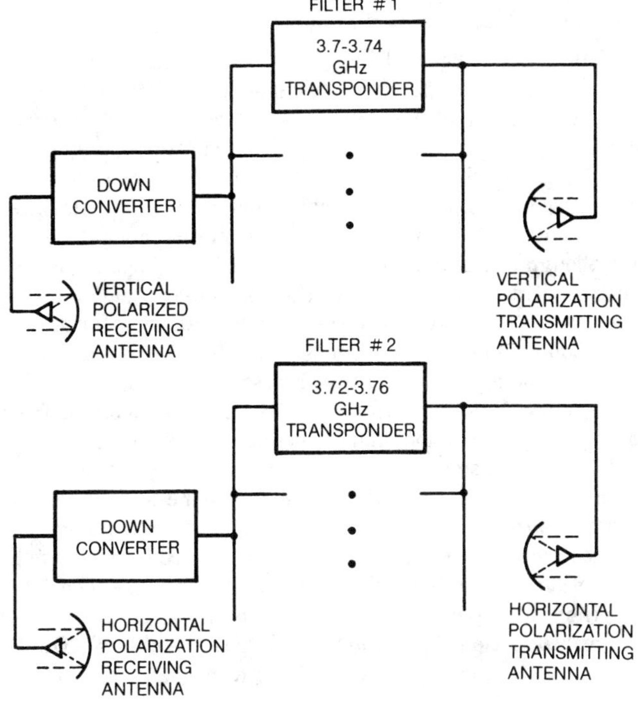

Figure 10-19. *Receiver Structure for Polarized Radiation.*

Multiple Bands

Another way to increase the number of simultaneous communications within a given satellite is to transmit and receive on several of the different bands of *Figure 10-17*. Communications in the 14/12-GHz bands can occur at the same time as communications in the 6/4-GHz bands, at least doubling the capacity of a given satellite. This approach has been used in the Intelsat 5 unit *(Figure 10-12)*. The problem with this technique is that the carrier frequency signals above 10 GHz are heavily attenuated by water droplets in intense rain, thick fog or heavy cloud cover. A suggested solution is to provide alternate earth stations separated by a distance of several miles. It would be unlikely that both would be blocked from the satellite view at the same time. This approach is called space diversity.

Modulation and Multiplexing Techniques

Another way to share the circuitry of a satellite is through multiplexing. Most of the early satellites had a frequency division multiplexing and many had the multiple access of FDMA. As the use of digital communications increases, time division multiplexing (TDM) is taking over. For TDM, shown in *Figure 10-19*, all stations use the same single carrier frequency, but their signal occurs at a specific time interval in a specific time slot. As in FDM, the time slot assignment could be permanent, though it is more likely to be assigned as the demand occurs. A central control notifies the members of the transmission path which time slot has been assigned, and the receivers only look at the information at these specific times. When the transmission is over, the time slot is released back to the available pool. This multiple access approach is called TDMA. TDMA offers some advantages over FDMA in that only a single frequency is used for a station connection, and the connection can be made through digital switches. In the specific example of *Figure 10-20*, station A has been assigned the first time slot over which it sends the message 1001. Station B has been assigned the second time slot over which it sends the message 0110. Both messages are to be sent to station C. The control system has assigned the other time slots to other communication tasks. The satellite multiplexes the information from A and B into the correct time slots onto the carrier sent to station C. Station C receives the combined TDMA signal and separates the two messages and deals with them as separate communications.

In *Figure 10-20*, individual bits from each of the data signals have been assigned time slots. It is also possible to assign a fixed set of bits such as an 8-bit byte or a 128-byte packet to a time frame, and time-multiplex these frames in a similar fashion. The satellite capacity has been increased significantly because several transmissions are sharing the same carrier frequency and transponder.

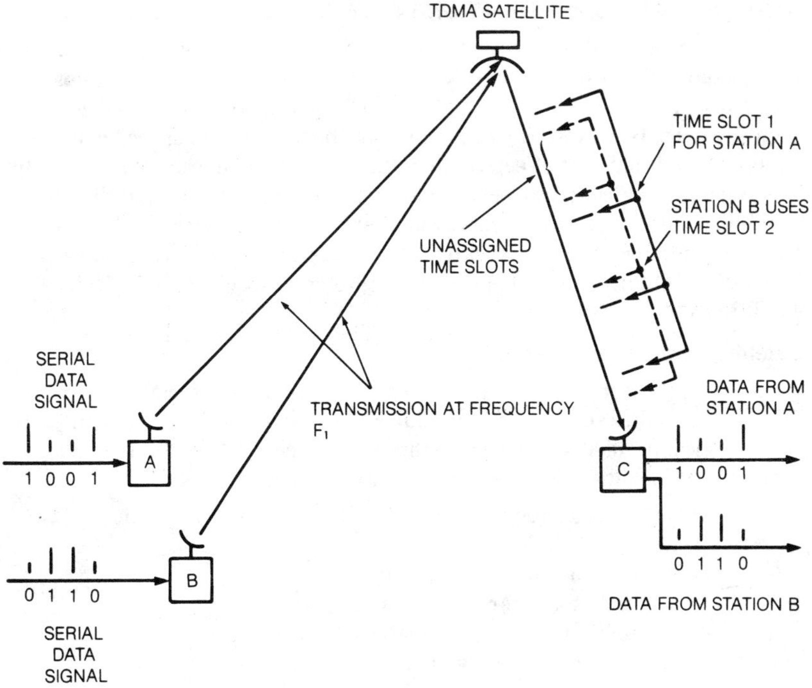

Figure 10-20. Basic Concept of Time-Division Multiplexed Satellite Communications.

Whatever the approach, the attempts to provide simultaneous utilization of the same channel of frequencies by several users usually complicates earth station equipment. Radiation polarization calls for the earth station to provide both polarizations in its antenna and power amplifiers. TDMA earth stations must be able to synchronize their digital transmissions with those to or from the satellite, and then demultiplex the signals. Large earth stations have intricate networks to accumulate information and distribute it over existing telephone or microwave relay networks to the destination.

As the satellite increases in size and capacity, it may also have the ability to communicate with other satellites directly instead of through a chain of earth stations. An interesting possibility for placing very large satellites in space may result from the availability of the re-usable space shuttle. Delivering satellites with it will allow satellite designers to provide heavier satellites of any desired weight (instead of only the discrete weights allowed by rocket launches) at a reasonable cost. This could be particularly beneficial to the small earth station user so he can become a much more active user of satellite communications systems in the future.

WHO CAN USE COMMUNICATIONS SATELLITES?

When satellites were first launched, only very large organizations such as the military or communications firms could utilize these systems. The overall costs were feasible only for the largest and most specialized of organizations. However, the units that have been placed in space have demonstrated that cost effective communications can be provided especially for multi-national corporations. As satellite power frequency and channel capability increase, it will be more and more feasible for small businesses and even individuals to utilize satellite systems. In fact, amateurs have been using satellites for the past decade through a special series of amateur satellite experiments.

Satellite Service for Individuals

While there may be some marketing problems associated with providing satellite services to individuals, there are not any insurmountable technological problems involved, as the amateur access to satellite communications has proved. Here is what is needed:

1) Integrated circuit receivers and transmitters using high-modulation index FM
2) Small fixed position antennas
3) Low-information rate transmissions
4) A satellite launched particularly for this service
5) All of the above provided at costs that are competitive with present communications, TV cable and entertainment centers.

It is felt that by launching a broadcast satellite using the Ku band that has sufficient broadcast power and enough bandwidth capability such that relatively narrow bandwidth transponders can be used, it would be feasible for individuals and small businesses to be able to participate in the system at a reasonable cost. Antennas of 1 meter to 3 meters (3 to 10 feet) in diameter seem sufficient for homes and businesses. Receivers that cost from 100 to 500 dollars would be adequate. It is felt that these costs can be considered as within the "reasonable" range.

If this system were activated, individual homes in remote areas could receive instructional material, entertainment, news, and sporting events over TV. Individuals in these locations could access data banks for library service, business service, electronic mail, or stock market quotations. Small businesses could carry on detailed business transactions that are not feasible at present.

Presently, TV programs from across the continents are brought into each home for occasions such as important international negotiations, the Olympic games, some earthshaking news report, etc. These are presently supplied by large communications companies. In the future these and many more types of communications will be received directly by individuals from satellites and due to satellite communications systems.

WHAT HAVE WE LEARNED?

- Conventional satellite communications use frequencies in the 4 to 6-Gigahertz band to provide communication between widely separated points on the earth's surface.
- A satellite in a geosynchronous orbit (about 22,700 miles [35,800km] above the equator) remains in the same position relative to the earth's surface.
- Satellites enable communication to occur between two stations, simultaneously between many station pairs, or from one station to many stations depending on how the satellite switches and handles the information.
- Satellites' capacities have continually increased since their inception in the early 1960's thanks to the improvements in rocketry, electronics (especially integrated circuits), communications techniques and antenna design.
- In the future satellites will make use of higher carrier frequencies, multiple frequency bands, dual radiation polarization, spot beams and advanced time and frequency multiplexing techniques to increase their capacity.
- Individual satellite earth stations using low-cost receivers and small diameter antennas are possible even with today's technology.

NOW FOR THE FUTURE

Large corporations, governmental and military organizations already use the entire range of communications techniques available. The thrust of the future will be to improve the techniques so that channel capacity is increased, reliability is improved, information rates are increased and costs are reduced. As these improvements are realized, the use of the techniques will spread to small businesses and into the home.

Homes of the future will be entirely different and lifestyles will be different. Inside the home there will be an information center connected to the external world by telephone, cable and broadcast television, radio and satellite. Information into the home will be provided by telephone, radio, television screen, computer CRT, a printer and possibly a facsimile. Besides the normal TV programming, expanded considerably by the cable and satellite links, the viewer will have the option of displaying on the screen at the same time as program material, explanatory material that provides background, clarification, language translation, etc. Information outputs from the home would be by telephone, computer keyboard, television camera, facsimile, and individual satellite links. With the home supplemented with a computer terminal and facsimile it is converted to a remote electronic classroom, or a remote electronic office for business purposes.

Business lifestyles could change considerably. Instead of workers using extensive amounts of energy to commute to central offices, they would stay at home, communicating with their fellow workers electronically with printed messages, computer data, interactive television conferences, and signals that control factory equipment.

Education of the family could change considerably. The members of a "class" would be located in their homes. Instructors would be connected to their class electronically, using the television screen for the personal contact; computers would be used to simulate problems, exercises, games, drill and practice and to test students. Massive computer files would provide complete library information to all students.

Home transactions would change considerably. Paycheck and bill payment transfers would be done by computer. A family member would shop by calling up store items on the TV screen, including different colors, styles, and price ranges. All order, delivery, billing, payments would be handled automatically by computer as each step is accomplished. Full family records for budgeting, income tax, investments, personal history would be kept by computer.

Homes would be automated to control supply and preparation of food, air conditioning and heating, and security systems. Fire and burglary alarms would be automatically sent to neighborhood safety stations for help. Medical aid emergency needs would be included with appropriate records on file for each family member.

Electronic information would saturate the home. Letters, newspapers, books, documents could all arrive electronically. Energy sources would be used for delivery of goods and services, recreation and social purpose. Commuting time would be reduced to a minimum and more time would be available for entertainment, social and service activities.

How fast these changes occur will be determined by the acceptability of the life style and the costs of the techniques. Through it all, communications – the systems and techniques – will help to improve the quality, safety and variety of life by providing the means for global interaction between individuals and societies through the transfer of meaningful information with ease and economy.

Quiz for Chapter 10

1. Satellite transponders are:
   a. a receiver and a transmitter designed to relay microwave transmissions from one point on earth to another.
   b. a device that echoes the radiation without change from one point on earth to another.
   c. devices that transform the message sent from one location on earth to a different code for transmission to another location.
2. Satellite transponders:
   a. use a single frequency for reception and retransmission of information from and to earth.
   b. use a lower frequency for reception of radiation from earth and a higher frequency for transmission of radiation to earth.
   c. use a higher frequency for reception of radiation from earth stations and a lower frequency for transmission to earth stations.
3. An up-link frequency for most satellites is:
   a. 4 GHz.
   b. 6 GHz.
   c. 10 GHz.
   d. 12 GHz.
4. A geosynchronous orbit occurs when:
   a. the satellite is placed in orbit 15,000 miles above earth in a north to south orbital path.
   b. the satellite is placed in orbit 22,000 miles from earth in an orbit that matches one of the longitudinal lines of the earth.
   c. the satellite is placed in orbit 35,000 miles from the earth in an orbit in the equatorial plane.
   d. the satellite is placed in orbit 22,000 miles from the earth in an orbit in the equatorial plane.
5. If a satellite is in geosynchronous orbit, it:
   a. remains in a fixed position relative to points on earth.
   b. can cover about 40% of the earth's surface.
   c. it moves faster than the earth's rotation so it can sweep across a large portion of the earth's surface in a 24 hour period.
   d. a and b above.
   e. b and c above.
6. Satellites allow earth stations to share the satellite equipment through:
   a. frequency-division multiplexing.
   b. time-division multiplexing.
   c. multiple frequency up and down links.
   d. focused spot beams and zone beams.
   e. all of the above.
7. The higher the frequency of the satellite or earth station transmitter:
   a. the higher the potential bandwidth that can be used for communications.
   b. the narrower the radiation beam for a given earth station or satellite antenna diameter.
   c. the closer satellites can be spaced without interfering with each other.
   d. All of the above.
   e. None of the above.
8. Satellite capacity depends on:
   a. the weight that can be placed in orbit.
   b. the panel area available for energy dissipation.
   c. the vacuum tube logic circuits and the power of the transmitter.
   d. All of the above.
9. Typical satellite weights that can be launched using DELTA or ATLAS/CENTAUR rockets in kilograms are:
   a. Less than 400.
   b. 400 to 1000.
   c. Above 1200.

10. A spin stabilized satellite:
- **a.** uses solar panels whose cells are continually oriented toward the sun.
- **b.** uses solar cells mounted on a cylindrical body that continually rotate so that about 40% of the cells receive solar radiation at a given time.
- **c.** uses gyroscopic action of a spinning satellite to maintain its orientation toward the earth and the sun.
- **d.** a and b above.
- **e.** b and c above.
- **f.** a and c above.

11. A body-stabilized satellite:
- **a.** uses solar panels whose cells are continually oriented toward the sun.
- **b.** uses solar cells mounted on a cylindrical body that continually rotates so that about 40% of the cells receive solar radiation at a given time.
- **c.** uses a spinning satellite to maintain orientation in space.
- **d.** uses a spinning momentum wheel to maintain orientation in space.
- **e.** a and c above.
- **f.** b and d above.
- **g.** a and d above.

12. Body-stabilized satellites generate _____ power than a spin-stablized satellite.
- **a.** more
- **b.** less
- **c.** about the same as

13. The typical satellite power in the 1970's in kilowatts is in the range of:
- **a.** 200 to 1000.
- **b.** 0.2 to 1.
- **c.** 0.5 to 1.5.

14. Satellite antennas of up to _____ feet in diameter have been successfully used in space.
- **a.** 5
- **b.** 10
- **c.** 15
- **d.** 30

15. Earth station antennas have been typically _____ feet in diameter.
- **a.** 25
- **b.** 50
- **c.** 90
- **d.** 150

16. For earth station antennas to be 6 feet in diameter, the satellite frequency bands must be in _____ GHz range.
- **a.** 4/6
- **b.** 12/14
- **c.** 20/30
- **d.** a and b above
- **e.** b and c above

17. Frequencies in the _____ GHz range are most affected by fog and precipitation.
- **a.** below 6
- **b.** 6 to 10
- **c.** above 10

18. As small antenna, low-cost earth stations become available, satellite communications can be used for:
- **a.** broadcast programs for information and entertainment.
- **b.** computer terminals in remote areas with access to large central computers with large information banks.
- **c.** emergency communications to remote areas.
- **d.** Any of the above.

19. In the future, satellite improvements will come from:
- **a.** increased power.
- **b.** switchable spot beams.
- **c.** time-division multiple access.
- **d.** frequency-division multiple access.
- **e.** All of the above.
- **f.** a, b, and c above.

20. Satellites may re-use the same frequency in the same area by:
- **a.** having many small antennas.
- **b.** overlapping radiation zones.
- **c.** TDMA.
- **d.** FDMA.
- **e.** Any of the above.
- **f.** c and d above.

(Answers in back of the book)

Glossary

A/D converter: A circuit that converts signals from analog form to digital form.

Address (Address code): A series of bits that denotes the location of information in a computer memory.

AGC (Automatic Gain Control): A control input to an amplifier that is used to control its gain, usually to keep the amplifier output constant as the input signal amplitude varies.

AM (Amplitude Modulation): A technique for sending information as patterns of amplitude variations of a carrier sinusoid.

AM demodulator: An electronic circuit that rejects the carrier frequency and recovers the modulation signal.

Amatuer radio operator: A private citizen who operates electronic communications equipment as a hobby.

Amplifier: An electronic device used to increase signal power or amplitude.

Analog: Information represented by continuous and smoothly varying signal amplitude or frequency over a certain range, such as in human speech or music.

Antenna: A device for radiating and receiving electromagnetic waves.

Aspect ratio: The width to height ratio of a two-dimensional image such as a television picture.

Asynchronous: Refers to circuitry and operations without common timing (clock) signals.

Bandpass amplifier: A circuit that increases the amplitude of signals whose frequencies fall between an upper and lower limit.

Bandwidth: The range of signal frequencies that a circuit or network will respond to or pass.

Baud: A unit of signalling speed equal to the number of signal events per second.

Binary Code: A pattern of binary digits (0 and 1) used to represent information such as instructions or numbers.

Bit: The smallest possible piece of binary information. A specification of one of two possible alternatives.

Body-stabilized satellite: A satellite whose solar panels are fixed toward the sun while antenna and body stabilization is provided through a spinning wheel.

Byte: A group of eight binary bits operated on as a unit.

Cable TV: A system for distributing television programming by a cable network rather than by broadcasting electromagnetic radiation.

Carrier: A signal which is modulated with the information to be transmitted.

Central Office: A local telephone switching center that provides connections to 10,000 subscribers or telephones.

Chrominance: That portion of a television signal that contains the color information of the picture.

Circuit: An interconnect of electrical components or devices that will perform some electronic function.

Citizen Band: The range of frequencies from 26.96 MHz to 27.41 MHz allocated to private citizen, short-range radio communication.

Coax Cable: A cable consisting of an inner wire surrounded by an insulating material, then covered by a metal shield and another insulating layer. It is used for transmitting frequencies in the VHF range and lower.

Communication: The transfer of meaningful information from one location to another.

Comparator: A circuit that compares two analog signals and provides a digital output that indicates which signal is larger.

Computer: The combination of a central processor, input/output and memory for storing and processing data.

Crossbar system: A matrix of switches used to connect one of a row of wires to one of a column of wires.

CRT (Cathode Ray Tube): An electronic device similar to a TV screen that provides a visual display output of stored or transmitted information.

Cuneiform: A form of written communications based on various patterns of triangles developed by the Babylonians.

Current: The flow of electrical charge, measured in amperes.

Cut-off frequency: The frequency above which or below which signals are blocked by a circuit or network.

D/A converter: A circuit that converts signals from digital form to analog form.

Data: Information about the physical state or properties of a system.

DC amplifier: A circuit that amplies signals with frequencies from 0 to a high cut-off limit.

Decoder: Any device which modifies transmitted information to a form which can be understood by the receiver.

Demodulation: The process of extracting transmitted information from a carrier signal.

Demultiplexer: A circuit that distributes an input signal to a selected output line (with more than one output line available).

Digital: Information in discrete or quantized form; not continuous.

Directional Radiation: Radiation that is controlled or concentrated by an antenna or focusing device such that it is received only by certain receivers.

Distributed Computer Network: A system whose computing or processing power is distributed among several computers.

Down-link: The carrier used by satellites to transmit information to earth stations.

EHF (Extremely-High Frequency): The portion of the electromagnetic spectrum from 30,000 megahertz to 300,000 megahertz.

Electromagnetic spectrum: The entire available range of sinusoidal electrical signal frequencies.

Electrode Type Reproduction: A facsimile process using special paper that is sensitive to electrical current or heat and turns dark in direct proportion to the amount of current introduced at various points on the paper.

Electronic Communication: The use of electrical signals to send and receive information.

Electrostatic Reproduction: A facsimile reproduction system which causes a powder to stick to paper by applying an electrostatic charge to the paper in the desired areas.

Encoder: Any device which modifies information into the desired pattern or form for a specific method of transmission.

ESS (Electronic Switching System): A computer controlled information connection or distribution system.

Facsimile: A method of transmitting pictures, printed pages, or film to a remote location where the information is reproduced in hard copy form.

FCC (Federal Communications Commission): A government agency that regulates and monitors the domestic use of the electromagnetic spectrum for communications.

FDMA (Frequency Division Multiple Access): The technique of allocating frequency multiplexed communication channels to satisfy user demands.

Fiber Optics: The process of transmitting infrared and visible light frequencies through a low-loss glass fiber with a transmitting laser or LED.

Field: Every other scan line in a frame of 525 scan lines making up a transmitted or received television picture – two fields per frame; 262½ lines per field.

Flyback pulse: The pulse in a television or facsimile signal that causes the scanning mechanism to return rapidly to the left side of the picture.

FM (Frequency Modulation): A technique for sending information as patterns of frequency variations of a carrier signal.

FM discriminator: A circuit that converts signal frequency variations to corresponding amplitude variations, i.e., a device that demodulates FM signals.

Frame: A set of 525 electron scan lines that completes the entire image of an electronically transmitted or received television picture.

Frequency: The rate in hertz at which a signal pattern is repeated.

Frequency-Multiplexed System: A communication system serving several users whereby each user is assigned a different frequency or bandwidth of frequencies within a specified range.

Geosynchronous Orbit: An orbit of a satellite that matches the earth's rotation so that any spot in the coverage area remains in the same relative position with respect to the satellite.

Ground Station: An installation specially constructed for communications with a satellite.

Hamming Code: A code that checks and corrects errors in binary information by comparing parity of received information with transmitted parity information.

HF (high frequency): That portion of the electromagnetic spectrum from 3 MHz to 30 MHz.

Hieroglyphics: A form of written communication developed by the Egyptians using a series of drawn pictures.

High-pass amplifier: A circuit that increases the amplitude of signals whose frequencies are above some cut-off frequency.

IF (Intermediate Frequency): Referring to bandpass amplifiers in superheterodyne receivers that are made to provide high gain and stable operation over a fixed range of frequencies.

Information: A physical pattern that has been assigned a unique and commonly understood meaning.

Ink-Jet Printing: A printing process which sprays ink onto paper through a nozzle with movement and pressure controlled by coded electrical signals.

Instruction Code: Digital information that represents an instruction to be performed by a computer.

Integrated circuit: A circuit whose connections and components are fabricated into one integrated structure on a certain material such as silicon.

Ionosphere: A layer of electrically charged particles at the top of the earth's atmosphere.

Low-pass amplifier: A circuit that increases the amplitude of signals whose frequencies are below a certain cut-off frequency.

Luminance: That portion of a television signal that contains the gray level information of the picture.

Mechanical Dialer: A method of selecting a specific numbered telephone using a rotating switch to interrupt the dc voltage a number of pulses equal to the desired number.

Microwaves: All frequencies in the electromagnetic spectrum above one billion hertz (1 gigahertz).

Modulation: The process of controlling the properties of a carrier signal so that it contains the information patterns to be transmitted.

Morse Code: A code developed by Samuel Morse which uses patterns of long and short pulses of energy to represent letters of the alphabet.

NAND gate: A binary digital building-block whose output is a 1 if any of its inputs are a 0.

Network switching: The method of electrical selection and connection of various available paths for transmitting information; usually for the telephones.

Noise: Any unwanted signal not present in the original transmitted information.

Oscillator: An electronic device used to produce repeating signals of a given frequency and amplitude.

PABX: A private automatic branch telephone exchange system used by individual businesses or governmental agencies.

Packet Switching: Computer communications of fixed length containing routing and distribution information as a part of the packet sent as a unit.

Parallel data: The transfer of data simultaneously over two or more wires or transmission links.

Parity: A binary bit that indicates the number of "ones" in a given binary code.

PCM (Pulse Code Modulation): A communications systems technique of coding signals with binary digital codes to carry the information.

PDM (Pulse Duration Modulation): See PWM.

Period: The time between successive similar points of a repetitive signal.

Phase: The time or angle a signal is delayed with respect to some reference position.

Phosphor: A material that emits visible light when excited by an electron beam.

Photodetector: A device that converts light intensity variations into corresponding electrical current variations.

Picture Element (Pixel or PEL): The smallest area of a picture whose light characteristics are converted to an equivalent electrical current or voltage.

Picture Tube: An electronic vacuum tube containing the elements necessary to convert transmitted electronic signals into visual images.

Primary Colors: A set of colors that can be combined to produce any desired color of the rainbow.

Processor: The central control element of a computer that obtains and executes the instructions contained in the program memory.

Program: The sequence of instructions stored in the computer memory.

PWM (Pulse Width Modulation): A technique for sending information as patterns of the width of carrier pulses, also called pulse duration modulation or PDM.

Radiation Patterns: The distribution in space of the electromagnetic energy produced by an antenna.

Raster: The scanning pattern used in reproducing television or facsimile images.

Rate of Information Transfer: The amount of information that can be communicated from sender to receiver in a given length of time.

Receiver: The person or device to which information is sent over a communication link.

Register: A series of identical circuits placed side by side that are able to store digital information.

Resolution: A measure of the smallest detail that can be seen in a reproduced picture.

RF (Radio Frequency): As opposed to sound or light or infrared or ultraviolet frequency.

Satellite Communication Systems: A remote communications technique using a satellite in orbit to receive signals from one location, and retransmit that information to another location.

Scan Line: One pass of an electron beam across the face or target of a television picture tube or camera.

Scan Pattern: The path of an electron beam which converts an image into electronic signals.

Serial data: The transfer of data over a single wire in a sequential pattern.

SHF (Super-High Frequency): The portion of the electromagnetic spectrum from 3,000 megahertz to 30,000 megahertz.

Signal Converter: A communications circuit that converts one form of information signal input into another form of signal output.

Smart Terminal: A computer terminal that has the additional capability to do local computation and processing of data.

Solar Panel: Specially constructed silicon panels which convert solar energy into electrical energy.

Sound Spectrum: The range of frequencies in the electromagnetic spectrum that can be heard by the human ear. Usually from about 20 cycles per second to 10,000 hertz.

Spin-Stabilized Satellite: A satellite whose antenna is kept stationary and on target while the satellite body containing the solar cells is rotated to face the sun.

Spotlighting: The process of focusing the light source on only the information to be reproduced in a facsimile system.

Spot Scanning: A photodetector that senses the light reflected from a small area is moved in a regular pattern over all portions of a page to be reproduced for a facsimile system.

Step-by-Step System: A telephone system that uses stepping switches or stepping relays to provide the communications path from calling telephone to called telephone.

Synchronous: Refers to two or more things made to happen in a system at the same time, by means of a common clock signal.

Telegraph: A digital system of electronic communication using short and long duration electrical pulses in a pre-determined pattern to represent letters of the alphabet. Invented by Samuel Morse in 1844.

Time Division Multiplexing: A communication system technique that separates information from channel inputs and places them on a carrier frequency in specific positions in time.

Time-Multiplexed System: A communication system serving several users whereby each communicator takes his turn during a specific period of time.

Toll Center: A major telephone distribution center that distributes calls from one major metropolitan area to another.

Touch-tone: An electronic method of selecting a specific numbered telephone by using buttons which activate different frequency tones for the desired numbers.

Transmission Link: The path over which information flows from sender to receiver.

Transmitter: The person or device that is sending information over a communication link.

UHF (Ultra-High Frequency): The portion of the electromagnetic spectrum from 300 to 3,000 megahertz.

Up-Link: The carrier used by earth stations to transmit information to a satellite.

VHF (Very-High Frequency): The portion of the electromagnetic spectrum from 30 to 300 megahertz.

Video Signal: The electrical signal containing the picture content information in a television or facsimile system.

Voltage: A measure of the electrical force that causes current flow in a circuit.

VCO (Voltage Controlled Oscillator): An oscillator whose frequency is determined by a control voltage.

Waveguide: A specially constructed metallic pipe for contained transmission of microwave electromagnetic radiation.

Wavelength: The distance between successive "peaks" of an energy sinusoid traveling at the speed of light (300,000,000 meters per second).

Index

Answers to Quizzes

Chapter 1

1. **d**
2. **d**
3. **a — D**
   b — E
   c — B
   d — A
   e — C
   f — F
4. **f**
5. **b**
6. **a**
7. **d**
8. **b**
9. **e**
10. **d**

Chapter 2

1. **b**
2. **a**
3. **b**
4. **c**
5. **d**
6. **b**
7. **a**
8. **c**
9. **e**
10. **b**

Chapter 3

1. **f**
2. **c**
3. **b**
4. **e**
5. **d**
6. **d**
7. **e**
8. **e**
9. **e**
10. **c**
11. **b**
12. **a**
13. **d**
14. **c**
15. **d**
16. **f**
17. **d**
18. **g**
19. **f**
20. **c**

Chapter 4

1. **c**
2. **f**
3. **d**
4. **g**
5. **b**
6. **d**
7. **d**
8. **b**
9. **d**
10. **d**

Chapter 5

1. **b**
2. **d**
3. **e**
4. **e**
5. **b**
6. **c**
7. **c**
8. **a**
9. **b**
10. **b**

Chapter 6

1. **e**
2. **d**
3. **d**
4. **c**
5. **d**
6. **d**
7. **b**
8. **d**
9. **c**
10. **b**

Chapter 7

1. **a**
2. **d**
3. **b**
4. **b**
5. **b**
6. **b**
7. **d**
8. **e**
9. **c**
10. **c**
11. **b**
12. **c**
13. **d**
14. **d**
15. **c**
16. **d**
17. **b**
18. **c**
19. **e**
20. **e**

Chapter 8

1. **d**
2. **b**
3. **c**
4. **a**
5. **e**
6. **b**
7. **a**
8. **d**
9. **d**
10. **c**

Chapter 9

1. **c**
2. **f**
3. **c**
4. **c**
5. **c**
6. **c**
7. **a. A, D**
   b. A, D
   c. A, C, E
   d. B, C
   e. B, C, D, E
8. **f**
9. **b**
10. **c**

Chapter 10

1. **a**
2. **c**
3. **b**
4. **d**
5. **d**
6. **e**
7. **d**
8. **d**
9. **b**
10. **e**
11. **g**
12. **a**
13. **b**
14. **d**
15. **c**
16. **b**
17. **c**
18. **d**
19. **e**
20. **f**